A GENTLE PLEA FOR CHAOS

A Gentle Plea for Chaos

Reflections from an English Garden

Mirabel Osler

BLOOMSBURY

For Eve Auchincloss

First published in Great Britain 1989
This paperback edition published 1995
Bloomsbury Publishing PLC, 2 Soho Square, London W1V 6HB

'A Word about Havoc' (Chapter 1) has appeared in *Hortus* under the title 'A Gentle Plea for Chaos', and 'Out Back' (Chapter 5) under the title 'Back Gardens'

A CIP catalogue record for this book
is available from the British Library

ISBN 0-7475-2120-4

3 5 7 9 10 8 6 4

PICTURE CREDITS
Jerry Harpur: pages 13, 33, 56-7, 84, 96, 101, 120-1, 125, 132, 157, 164, 169
Michael Osler: pages 20, 24-25, 29, 40-1, 49, 64, 69, 77, 88-9, 109, 137, 141, 148, 152-3

Engravings by Howard Phipps
Designed by Fielding Rowinski
Typeset by Bookworm Typesetting, Manchester
Printed in Hong Kong

Contents

1 A Compulsion for Trees

Origins of the Book / The Far East, Mogul and Ottoman Gardens / A Glance at Italy and France / A Word about Havoc / Our Garden / Planting Trees / The Enchantment of Trees

When the ice of winter holds the house in its rigid grip, when curtains are drawn early against that vast frozen waste of landscape, almost like a hibernating hedgehog I relish the security of being withdrawn from all that summer ferment that is long since past. Then is the time for reappraisal: to spread out, limp and receptive, and let garden thoughts rise to the surface. They emerge from some deep source of stillness which the very fact of winter has released.

This is the alchemy of gardens. For I have come to understand that gardening is not just a putting into the earth of some frail greenery, but like a stone thrown into a pond, garden thoughts ripple outwards towards a limitless horizon. What are gardens? Who are gardeners? Where does the thrust to make places of beauty, secrecy or seclusion come from? Who were the first? And what intricacies are we led into from the sheer chance of planting a seed in the earth?

So I imagine up and down the country during these blessed months of short days and long nights, a whole self-seeding of gardeners, with backs unbent, having put aside their boots, trowels and twine, who can now have time to let their thoughts hang out: a time when everything is possible. Who doesn't make lists then? Heady, wild and totally outrageous ideas can be brought into line, maybe only momentarily, before they are banished as unrealistic. It is the season for minds sharp as blades – agile and springing from one extravagant thought to another.

You don't have to garden to garden; gardening in the mind is a gentle vice with an impetus of its own; it may not be as potent as actually making one, but there is a whole different threshold where gardening in the head can fill our winter tranquillity with unrest. What gardener can condemn this as a time of stagnation?

Books and catalogues come into their own in winter. Not only those on *How To*, but those filled with plans and pictures; books by inspired gardeners who have been at it for years and reassure us of how easy it is, and who lead us through labyrinths of scented borders, classic design, *potagers* and pleached *allées*. Or pick up one of those superb books such as *Private Gardens of England*, by Penelope Hobhouse, on a wild, wet afternoon in February when the wind is shrilling outside, moaning through the gaps and spattering the window with rain. Turn to a picture of Saling Hall showing blues and silvers against the static severity of evergreens, or see the black and white photograph of roses, cobbles and fennel at Chilcombe, in which there's a table and chair glimpsed through an open door in the garden wall. Books such as these are indeed a strong element of the whole pleasure of gardening; they need to be devoured and mulled over as well as those which are carried round in an earthy hand as vital advice flows out on what to do with five hundred slugs at the delphiniums.

Each year the choice of gardening books in print seems to multiply from those with such evocative titles as *Plant Hunting on the Edge of the World* to *A Brocade Pillow – Azaleas of Old Japan*. Intriguing as they sound I know I shall never be aroused by *Succulents*, neither will I *Know Your Woody Plant*.

This book does not offer helpful advice, rather it is a collection of ideas originating from making a garden, some of which perhaps are shared by others who, in winter, or on their knees in summer, have time to ruminate on the infinity of gardens. In a way this is the anti-gardening gardeners' book. Yet that's too glib, too simplified. What I want to do is to let in a cold blast of high altitude air to make some gardeners gasp from either indignation or pique or, better still, agreement. These garden thoughts have sprouted while I've been dead-heading roses, visiting gardens or buying a pair of socks. No matter from where, the creativeness of gardens must have plagued men and women for centuries. But trying to tie down such random ideas is like trying to constrain that wayward clematis, 'Etoile Violette', with her capricious tendrils which reach everywhere.

There are many fascinating or rarefied aspects that I have not pursued though I'd like to, such as secret gardens, plant fossils, John Evelyn or orangeries; others are too academic or scholarly: medicinal plants and the whole wide territory of garden design. Instead I have taken the five major aspects of our own garden which have had the most forceful effect on its layout. I have divided the book into chapters: Trees, Water, Walls, Roses and Bulbs. The themes around these five subjects are arbitrary; their sole reason for being included is that they have each originated since 1980 when we first began to make the garden. Without that incentive not a single garden thought would have infiltrated my mind.

Garden concepts are threaded through centuries: we pull on them as on a string, not knowing from where our inspiration germinated, only that we each gather up different threads to form whatever pleases us. We know what weight and emphasis we want. And when we walk through other people's gardens there are places where we recognize instantly that their inspiration is absolute. It almost sings with its rightness.

Where did gardening begin? Putting Genesis aside, was it in the Old Kingdom of Egypt, with its pictures of papyrus and lotus? Or was it in the Hanging Gardens of Babylon, watered by the Euphrates; with the sepulchral lilies of Santorini; in Ancient Greece where Homer sang and the blessed lay decked with asphodels amongst the Elysian Fields, or in the Hesperides, those fabled gardens at the edge of the western world, where golden apples grew in the Isles of the Blest? Gardens wind back through sacred groves, sanctuaries and arbours; through Roman or Islamic gardens brought by the Moors to Spain, where formal, grassless courtyards, arcaded and with tiled seats, contained a variety of citrus trees.

What were Far Eastern gardens like? Their development over a thousand years may seem alien and, without having seen them, I can only guess by looking at pictures, and yet I find their pared appearance, the spareness and immense skies, appealing in contrast to our own English gardens. Where ours may sometimes look romantic or rumpled as well as formal, here in the East, is deliberation and symbolism from a single pine tree or rock, from a snowy mountain or merely a branch of plum blossom. In a painting the arrival of spring is barely implied by the mist rising to disclose fresh growth.

Chinese gardens were devised to create a continuous flow leading on from one place to another: an imaginative journey where paths led through doorways, along walls or across bridges, with the continual promise of further revelation. Water was used to make 'lakes', rocks to make 'mountains', and symmetry was not a prerequisite. Instead there was the contrived contrast from yield to rigidity, sunlight to shadow, closeness to distance and hidden to revealed – every aspect had its opposite; each was there to create the yin and yang elements which are at the core of Chinese gardens. Plants were used not for their rarity or beauty, but for their symbolism. Bamboo, which bends in the wind, infers an honourable man, the lotus is the rising soul, the peony wealth, the peach fecundity and so on.

Look at a Chinese painting of a tree peony and a quince to see how sensual are the soft and brilliant colours; how spontaneous the brushwork appears and how, above all, the transient delicacy of what had inspired the painter still reaches us more than four hundred years later. See how in a picture or scroll there might be just a glimpse of roof or bridge, pine tree or thicket of bamboo. In place of lawns, there is water, providing the tranquil space where reflections make it a place for contemplation and serenity. Unlike the Mogul or Ottoman gardens which were retreats from the wilderness, Chinese gardens were havens from the ordered austerity of daily routine: leaves fallen from a tree were left in their significant circle, reflecting the outline of the branches.

Japanese gardens, although influenced by China, were earlier characterized by their cult of Shinto, which delighted in the individual beauty of an ancient tree, a waterfall or a rock. Divinity was perceived in every natural object; nature was revered and Shinto shrines grew out of the landscape with natural perfection. The *Tale of Genjii*, a tenth-century novel by Lady Murasaki Shikibu, tells of gardens which were made for the many women that Genjii loved, and the novel inspired many artists to illustrate it. These artists have influenced succeeding garden designs, including one south of Kyoto, where a platform for gazing at the moon beside a lake was meticulously recreated to evoke scenes from the *Tale of Genjii*, seven hundred years after the book was written. Another garden influence was the tea ceremony, which took place in a small rustic hut with tatami mats on the floor, where there was just enough room for a few friends to enjoy a ritual bowl of tea together. Outside there would be a traditional stone basin for washing hands, stepping stones to reach the hut, and even a stone lantern to light the way after nightfall.

In Japan there are times of day, or times of year when deliberate viewing of cherry blossom or the melting snow takes place. I wonder how this would work in our gardens? For though we know that there will come a certain day when something is at its best, contemplation of it has to be an almost solitary affair, because you can never tell beforehand what the weather will be like, what damage will be done or even how eccentric the seasons may be that year. Who hasn't stood in their garden at some unexpected moment of the day, when perhaps the tension in the petals of a tree peony is almost a breath away from dissolving, or when the immaculate clarity of a tender arum lily seems becalmed for a moment before the petal curls too emphatically? Or when in a certain light there is an almost smoky aura given off by the mauve and white Japanese anemones, when black thunder clouds pass behind a laburnum tree in full flower, or when frost outlines a head of winter yarrow – how often then have we wanted to share it? But try to organize a midsummer flower party when scents and blooms are filling the garden and you can be sure it won't come off. So how did they manage in Japan? Were seasons fixed events? Did things work according to the time of year and did weather behave seasonally? Looking at the reticence in Japanese paintings, with their one exquisite branch or plant, I suppose this was one way out of the problem.

In complete contrast to the Far East are the enclosed and formal Mogul and Ottoman gardens. The former were once such magnificent places, with their four divisions of the *charbagh* – symbols of the elements: earth, air, water and fire. What inspired havens of serenity and refuge the six great Mogul emperors created, especially in Rajasthan, where the heat was relentless and the dry, stony hills were waterless. The apotheosis of these vernal illusions, all green and flowery with their calm designs and shady trees, began in 1525 and lasted for almost two hundred years. Coming from a land of water, mountains and orchards we can imagine how those emperors were desperate to create their own bearable sanctuaries. With the use of water in straight canals flanked by cypress trees, with parterres filled with flower patterns resembling fine embroidery, and with small open-sided pavilions on platforms constructed to catch whatever passing breeze there was, they surrounded themselves with watery sounds from plumes or small jets which cooled the air. Water rippled continuously down scalloped chutes. Rose and lily, tulip and iris provided imaginative retreats on a small intimate scale in the midst of such a pitiless landscape.

Nowadays walking through these sad gardens, all has long since vanished; nothing is left but the geometric designs, with their precision of patterns from which vitality and warmth have disappeared. Such shabby places, forlorn and neglected; water no longer flows, flowers are meagre and trees that once supplied their essential benison of shade have gone. What we read about we have to imagine; in our mind's eye we recreate that care and detail. It needs a great leap of the imagination to revitalize the gardens with colour and animation, to visualize the feasting and coloured rugs, the dishes and embroideries, and the russet, azure and gold of the costumes of the harem women. Gone are the sounds of soothing water and musicians; now only the piercing cries from flocks of viridian parakeets and the clicking of cameras are to be heard.

Ottoman gardens, like those of the Moguls, were intimate. By laying a rug in the wilderness, the first gardens were created by nomadic Turks. Later, when they cultivated flowers, it naturally followed that in such a waterless land, trees should be planted to shade the flowers; walls were then needed to protect the trees and more walls to protect the harem. Within these sanctuaries water was also used for its assuaging sound as it sprang from small fountains, or for its gentle motion, continually falling from scalloped basin to basin. Beds of scented flowers and roses, twining climbers, carvings and scrolls all created that same floral privacy as in the Mogul gardens. And there was one distinctive feature of the Ottoman garden – the kiosk – a small, roofed pavilion often decorated with mother-of-pearl and gilding, and hung with paintings of fruit and flowers. Descriptions of these kiosks from travellers in other centuries sound too delectable, with their places for reclining and their splashing water from interior marble fountains. Contemplation and inactivity could last for hours. Imagine. There is no doubt that just to read about or to look at pictures of them, is to be aware of an immediate and enviable charm that is intrinsically their own. One legacy from sixteenth-century Turkey, for which we must be forever thankful, is the tulip. What a debt we owe to Suleiman I for taming the tulip, the flower of his court, depicted so ravishingly on innumerable tiles and pots, and in illustrations. How much more trenchant a legacy than bonsais from Japan.

Developing almost simultaneously, give or take a hundred years, were the great gardens of the Renaissance. Here was design on an immense scale, where gardens were integrated with their villas, making a complete extension to the building. They broke away from earlier enclosures such as at Il Trebbio, the hunting lodge of the Villa Medici in Tuscany, where the medieval garden is still in existence: inward-looking and sheltering. No use was made of the wide views from the hill top. Yet within a few years everything changed. Spacious terraces, wide walks leading to pavilions and

magnificent views, which gave the visitor the genuine feeling of 'beholding', formed a unity of house, garden and countryside.

I find it hard not to be intimidated by looking at the designs and paintings of such places as the Villa d'Este at Tivoli, or Vaux-le-Vicomte near Melun, in France. Here again, as in the Mogul gardens, is that quality of detachment, of being uninvolved in the gardens. For those of us conditioned and brought up not to take a historical view of gardens, or who haven't the architectural and retrospective grounding needed to come to terms immediately with design on this scale, we are left unable to assimilate, or even to focus on what we are seeing. What *are* we seeing? Is it brilliance in regal proportions, or is it the natural progression of garden layout passing through an imperial epoch? How do you absorb such scale? It's a far cry from the homage to nature sung by Boccaccio and Petrarch. And worlds away from our own cosy concepts of garden schemes. But if you can wipe away any preconceived ideas and look with a mind free of imprint, there is something compelling and strong in design on such a scale: in the dignity from identically clipped and precisely placed box and yew blocks; from the drama and boldness of perspective. There is an even more recognizable response from the partially eroded sculptures of figures, creatures or fountains as they have weathered over the generations. Looking at Le Nôtre's design for Versailles, it needs a particularly open and curious mind to see this place when the dimensions are of such implausible proportions. Versailles may have glorified the Sun King, but for the visitor with not much time to adjust, it is hard to know exactly what it is you are dealing with; it's as though you are walking within proportions that are completely unassimilable.

One last thought on France before leaving for England – why on earth do the French torture trees? Why those aggressive amputations? Espalier and cordon, fan-trained or pyramidal are all brilliant and desirable forms in certain places, but in France they seem to go mad with hostility towards their trees. Rows and avenues are decapitated, willows are shaved into regimental heads. Is it frustration with the shears between one vine pruning and the next, when a passion for lopping, snipping, slashing or cleaving has grown into an addiction?

Returning to our gardens, to our own small plots, who really cares from where our zest for gardens originates? Whether from myth, hearsay or historical facts, from paradise or Persia, let's be less cerebral and more eclectically wanton. Merely thinking formless thoughts is enough to add another thrust to the whole gardening mystique.

Looking round gardens, how many of them could be said to lack that quality that adds an extra sensory dimension for the sake of orderliness? There is an antiseptic tidiness that characterizes a well-controlled gardener. And I'd go further and say that usually the gardener is male. Men seem more obsessed with order in the garden than women. They are preoccupied with flower bed edges cut with the precision of a pre-war hair cut. Using a lethal curved blade, they chop along the grass to make it conform to their schoolboy set squares, and with a dustpan and brush they collect one centimetre of wanton grass. Or, once they get hold of a hedge-trimmer, within seconds they have guillotined those tender little growths on hawthorn or honeysuckle hedges that add to the blurring and enchantment of a garden in early June.

The very soul of a garden is shrivelled by zealous regimentation. Off with their heads go the ferns, ladies' mantles or crane's bill. A mania for neatness, a lust for conformity – and away go atmosphere and sensuality. What is left? Earth between plants: the dreaded tedium of clumps of colour with earth in between. So the garden is reduced to merely a place of plants. Step – one, two. Stop – one, two. Look down (no need ever to look up, for there is no mystery ahead to draw you on), look down at each plant. Individually each is sublime, undoubtedly. For a plantsman this is heaven. But where is lure? And where, alas, is seduction and gooseflesh on the arms?

There is a place for precision, naturally. Architectural lines such as those from hedges, walls, paths or topiary are the bones of a garden. But it is the artist who then allows dishevelment and abandonment to evolve. People say gardening is the one occupation over which they have control. Fine. But why over-indulge oneself? Control is vital for the original design and form; and a ruthless strength of mind is essential when you have planted some hideous thing you lack the courage to demolish. But there is a point when your steadying hand should be lifted, and a bit of native vitality be allowed to take over.

One of the small delights of gardening, undramatic but recurring, is when phlox or columbines seed themselves in unplanned places. When trickles of creeping jenny soften stony outlines, or when Welsh poppies cram a corner with their brilliant cadmium yellow alongside the deep blue spires of Jacob's ladder, all arbitrarily seeding themselves like coloured smells about the place.

Cottage gardens used to have this quality. By their naturally evolved planting, brought about by the necessity of growing herbs and fruit trees, cabbages and gooseberries, amongst them there would be hollyhocks and honesty, campanulas and pinks. How rare now to see a real cottage garden. It is far more difficult to achieve than a contrived garden. It requires intuition, a genius for letting things have their head.

In the Mediterranean areas this can still be seen. Discarded cans once used for fetta cheese, olives or salt fish, are painted blue or white and stuffed to overflowing with geraniums placed with unaffected artlessness on steps or walls, under trees or on a window sill. Old tins are planted with basil, and stand on the threshold of a house, not for culinary use, because basil is a sacred plant, but for the aromatic pleasure when a sprig is picked for a departing traveller. Under a vine shading the well, are aubergines, melons, courgettes and a scatter of gaudy zinnias. An uncatalogued rose is grown for its scent near a seat over which a fig tree provides shade and fruit. Common sense and unselfconsciousness have brought this about. A natural instinct inspired by practical necessity. We are too clever by half. We read too many books, we make too many notes. We lie too long in the bath planning gardens. Have we lost our impulsive faculties? Have we lost that intuitive feel for the flow and rightness of things: our awareness of the dynamics of a garden in which things scatter where they please?

This brings me to another observation which I think goes with my original longing for a little shambles here and there. For it seems that proper gardeners never sit in their gardens. Dedicated and single-minded, the garden draws them into its embrace where their passions are never assuaged unless they are on their knees. But for us, the unserious, the improper people, who plant and drift, who prune and amble, we fritter away little dollops of time in sitting about our gardens. Benches for sunrise, seats for contemplation, resting perches for the pure sublimity of smelling the evening air or merely ruminating about a distant shrub. We are the unorthodox gardeners who feel no compulsion to pull out campion among the delphiniums; we can vacantly idle away small chunks of time without fretting about an outcrop of buttercups groping at the pulsatillas. Freedom to loll goes with random gardening, it goes with the modicum of chaos that I long to see here and there in more gardens.

Not all gardens fail in this, of course. There are two, for instance, that have this enchantment from the moment you enter. One belongs to people we know who live on the Welsh borders, where all the cottage attributes such as mulberry, quince and damson trees grow amongst a profusion of valerian and chives, marjoram and sedums. The whole lush effect is immediate and soothing; it gives you a feeling of coming home, it reminds you of what life ought to be like.

In complete contrast is Rosemary Verey's garden at Barnsley House, near Cirencester, in Gloucestershire. Here among the strong lines of design, parterres and walks, the classical temple and knot garden, one has the impression that the owner washed over the whole layout with soft, diffused colours so that hard lines became blurred. Sweet rocket and violas, rock roses and species tulips beguile, flow and confuse. It may not be chaos – it certainly isn't – but it is as if this truly cohesive effect happened while the owner had her head turned away. She hadn't, we know, for a garden like this was painstakingly achieved from the brilliance of deliberation: knowing when not to do things as vitally as knowing when to. There isn't a dandelion unaccounted for.

So when I make a plea for havoc, what would be lost? Merely the pristine appearance of a garden kept highly manicured, which could be squandered for amiable disorder. Just in some places. Just to give a pull at our primeval senses. A mild desire for amorphous confusion which will gently infiltrate and, given time, will one day set the garden singing.

There are many ways of starting a garden. Abstract ideas may originate in the mind, and are then meticulously transferred to paper, when every bed is plotted for colour, shape and size before the first plant goes in. Another impetus may come from a cherished longing to have one area of your own, where no one can constrain you and where no conformity compromises your imagination. Or, after the culmination of years spent hoarding articles and seed catalogues, the gardener knows already what needs to be done. Yet others may be haunted by childhood memories of magic places of make-believe, of games, scents and secrets. Gardens may start from a bare piece of earth surrounding a newly-built house, or from the sheer necessity of hiding some hideous building, or maybe from a desire for self-protection from sea and tempests. Others need a little space for sitting in the sun, for hanging out the washing, where children can play or just where the cat naps. Whatever it is, once started a garden holds you in its thrall. However irksome it becomes at times, who can go outside and kick a lily?

How did we begin our garden? I don't know, it wasn't deliberated. My husband Michael and I knew nothing, and anyway we wanted freedom to travel. Yet our very landscape coerced us. With one and a half acres of undulating land, a winding stream, stone buildings and old orchard trees, how could we have resisted the temptation? We didn't, of course. One bulb put into the ground inadvertently, two or three trees planted here and there,

and we were hooked. By slow infiltration our garden began. Inevitably we started to plan, not grandly or methodically, with spaces worked out in inches, or with a basic design of what we should be about, but on scraps of paper as we sat over a meal or walked round with a notebook and pencil. Our ideas tumbled; mad or possible, they either took flight and vanished, or we held on to them arguing round their feasibility.

What did slowly emerge were five elements that were important in the garden. Like strong characteristics they influenced us in our decisions. Water and walls were the two existing attributes that led us naturally to take a certain course, and even now they have ascendency on what we do next. The source of our brook is in the large, rounded Shropshire hills which rise to eighteen hundred feet just behind the village; at one point this brook goes right through the village, flowing for some fifty yards along the road before linking many of the low-lying cottages and farms. Then our cottage, having once been a farm, has several outbuildings: a cowshed, a granary and a pigsty, as well as the old outside privy and miscellaneous garden walls. So water and walls are a most integral part of our garden; they are dominant features which persuasively turned us in particular directions.

The shape of the land placed an almost unconscious bias on our plant decisions. The slopes and steep banks forced a bossy geography on any planting outline. Surveying what was there left little doubt in our ignorant minds that bulbs would do, planted wildly and strewn about the landscape where shrub roses would also grow. These roses would be bountiful affairs dumped into the grass, loving it and looking spectacular. That was our intention at any rate. Capricious fancy, which may never be realized, is part of the pleasure of planning gardens.

Finally there were trees. We already had an upper orchard full of old, useless apple and pear trees, which still blossom unsparingly in spring, provide height and swarthy shadows in summer, and in winter become a massive tangle of brittle twigs; yet these trees always give a bulky presence at any time of the year. Trees were so inherent in our scheme that we never even questioned an alternative. Any garden must begin with them. They are the salient features around which everything else is worked.

Perhaps it seems a bit perverse or unorthodox to plan a garden without flowers, but that is just what we did. Trees were our first preoccupation. And once having started we didn't want to give up; not when what you do is so often barely visible; when you have to be patient and wait six years or so to see what you are up to. Build a wall, paint a picture, sew a fine seam, the result is instant. You stand back and admire or wring your hands in despair, but at least the outcome is tangible. But with planting – trees or bulbs – there is the catch and the wizardry; for who can turn their backs on their first spring blossoming?

Five *Prunus avium*, the ordinary woodland cherry, were our first, with two mountain ash, *Sorbus aucuparia*, and a walnut. The very business of choosing, siting and imagining them full-grown precipitates further acquisitions. For the next two years trees overwhelmed us. We have a six-acre field and into part of it, after fencing off a piece from grazing sheep and digging the wire netting deep enough to keep out the rabbits, we planted a mixture of thirty-five beeches, oaks, limes and chestnuts, interplanted with over a hundred larches to nurse them along. The words 'copse', 'spinney', 'woodland', 'grove', and 'brake', sounded so thrilling as they coursed through our minds, and were followed naturally by 'conservation', 'wild life' and 'precipitation' so that by the next year we hived off a larger piece of field – two acres – which we fenced and planted with more trees. This time it was five hundred and fifty hardwoods: more oaks, ashes and beeches, limes, planes and birches, a few Scots pines and wild cherries. Again we planted larches, eight hundred in between; these will eventually be taken out when the broad-leafed trees have outstripped the larches, which are such rapid growers. Probably in about ten years' time.

It hasn't all gone swimmingly. Not at all. The first group of thirty-five, being on a south-facing slope of the field have grown vigorously; in fact the advice we had been given turned out to be true: that nothing is gained by planting a twelve-foot tree instead of an eighteen-inch whip, for this really will catch up with the large tree in a few years. It happened. Within five years our beeches and oaks were just as tall as the twelve-foot chestnuts which barely moved at all in that time; the trauma of being transplanted had left them stunned.

But the following year's planting of hundreds of trees has not burst into enthusiastic growth at all. They too were put in as small whips, but because this time the slope is north-facing, they have grown with painful forbearance. Frost smote the ash tree buds as they were just coming into late leaf in May; two unspeakably wet winters bogged down the limes which hated the treatment; the southern beeches couldn't stand the cold, and rabbits ate the larches because it had been too expensive to fence them in, as we had done with the first small group.

It sounds dire, I know. But amongst this carnage most of the trees are surviving. It's just that we can't throw our hats in the air with confidence that they are thriving. However we only have another fifty or so years to wait, and then it will be possible to walk through a small woodland where the leaves underfoot are deep and shuffly, and the layers above are dense enough to make a dappled pattern on the trunks of the beech trees.

Meanwhile in a corner of the field we are slowly, very slowly – because we are doing it by hand – digging out an old silage pit which one day we hope will make an all-round-the-year pond for wild life. Already wild duck use it in winter, curlews call from the next field and we hope the fox knows our place is a safe haven from assassins. Unless you are very young it may seem an act of bravado to plant two hundred eighteen-inch oaks. 'Ah, yer'll never live to see 'un grow,' our morose old neighbour said as he saw Michael planting a sweet chestnut. But that was not the point, was it? Planting trees is a gesture into the future, it is a hand held out to other generations. We can't all build a Parthenon.

Trees embellish a landscape and a garden. Even in small places where size is restricted, the mere presence of just one tree like a twenty-five foot Snowy Mespilus, *Amelanchier laevis*, with white filigree blossom, instantly adds a serious dimension to the overall effect with its rounded head. The fallen leaves make a glowing vermilion scatter on the ground if they aren't swept up by an over-zealous gardener. The whole design of a garden and the proportion of the house, are immediately lifted by the vertical elevation of a tree or trees.

Through our slow assembling of knowledge, through mistakes and advice, one job remains constant – the planting of trees. Years after we first started our small wood we have gone on adding trees in the garden. Perhaps more than anything else we plant, this gives us the most satisfaction. We have spent many hours looking through tree books trying to gauge the reality of a twenty-, thirty- or forty-foot high tree in this or that number of years, for although we will not be around to see them mature, we are already discovering that some trees are planted too close to something else. How hard to hold that invisible dimension in space as we pace the ground looking for the right position. A weeping mulberry, *Morus alba* 'Pendula', (incidentally I wish we had bought *M. nigra*, the delicious Black Mulberry, instead of the boring white) is rashly near to a huge loaf of a shrub rose which like leavened dough keeps rising. The circumference of this Rugosa,

'Scabrosa', will, any day now, be pressing against the drooping branches of our young mulberry, barely seven feet tall, and I haven't yet the heart to cut back those magenta flowers and bright green leaves of the rose. But it is not always like this. Things have gone right. On an east-facing bank where the grass is roughly scythed once a year, we have planted a tree with the unbelievably poetic name of liquidambar, *L. styraciflua*, or, more prosaically, 'Sweet Gum'. It lives up to its name by not only having good maple-like foliage, but, in autumn, with luck, by turning into a radiant creation of crimson, purple and flame leaves. On the same bank are two maples, *Acer pensylvanicum*, with striped white and jade-green stems, like some expensive woven silk and *A. davidii* 'George Forrest', with very good red-stemmed large leaves. Nearby a malus, 'Golden Hornet', produces a lavish crop of deep yellow fruit which stays on the tree long after the leaves have fallen, and a sorbus, *S. aria* 'Majestica', a slightly more sober tree than the others. But a tree of singular beauty, if only it would get on with it, is a Davidia, *D. involucrata*. This has such a mass of evocative names that visions of transparent airiness immediately form in my mind's eye: the 'Pocket Handkerchief Tree', the 'Ghost Tree', the 'Dove Tree' and no doubt a whole lot more descriptive names. Its drama and translucent effect is to be seen in May, when the whole tree shimmers with large creamy bracts. If you can't swear for sure you know what a bract is, according to Hilliers' *Manual of Trees and Shrubs*, it is 'A modified, usually reduced leaf at the base of a flower-stalk, flower-cluster, or shoot.'

The vast family of prunus which includes almonds, peaches and apricots also provides us with some of the best garden trees. We planted a *P. serrula*, a small and safe choice with polished bark the colour of mahogany. But watch it with these trees. The many prunus from Japan can be pitfalls for the unwary. Some of the suffocating pink froth that is seen foaming in the suburbs against a red-brick background come into this category. We were not immune. Hoping for something with a generous white lushness to be seen languishing with a green bank beyond, we bought a prunus called 'Cheal's Weeping Cherry', although correctly it is called 'Kiku-shidare Sakura' - ah, if only we had read the Latin, *serrulata rosea*, we might have been more prudent. How out of place its showy deep pink flowers looked, reeking of the boudoir and the Ziegfeld Follies; how inhuman and brutal we were to dig it up and to banish something living so happily amongst the gentle primroses, the shrinking violets. But we were. We did find someone longing for just such a tree and now I hope its pink superfluity is throbbing

through the urban streets for miles around.

These are just a few of the trees we planted along this particular bank – altogether we put in over thirty, including four Irish yews. They are aristocrats amongst evergreens, at least in England, where we can find nothing with the handsome nobility of the Mediterranean cypress. *Taxus baccata* 'Fastigiata' or Irish Yew has a black-green density, strong and dominating, which often stands high above grey and mossy gravestones, or is seen flanking a path to the church. Here along our bank in the upper orchard, we want our yews to contrast with the flimsy limbs of flowering trees, with their grace and open growth.

Although I have described the fruit, berries and autumn colours of some of our trees, I can't say it comes from what I actually see in the garden. It really comes from what we saw when we visited the arboretum, as we made the first list of trees that we wanted. Westonbirt Arboretum in Gloucestershire is a place to visit for any tree addict; the variety and magnificence of everything you see there is impressive. It certainly inspires a new gardener to raise her head above flower bed level.

Impatiently we wait for these many slender trees, some of which are already covered with yellow, amber and carmine leaves, berries and fruit in autumn, to grow large enough to form shapes. To be able to walk *under* the branches of a tree that you have planted is really to feel you have arrived with your garden. So far we are on the way: we can now stand beside ours. Even so, in spring and autumn they draw us back to look, admire and to measure minutely as they reach shoulder, head or greater height year by year. There is no doubt that tree planting means taking a long view; nothing is instant, nothing is transitory, it has the basic powerful attraction of construction with a living component. In a landscape where all the traditional elms have been felled, it is not bad to stand back from a two-foot oak you have just planted as you visualize a ninety-foot high head of foliage which one day will cast its rustling shadow in a seventy-foot spread. For some mad reason, which I've long since forgotten, in an early year we planted a metasequoia. This is not a tree really suitable for here at all. (*Metasequoia glyptostroboides* is, for those who are interested, its proper Latin name; it is also known as Dawn Cypress or 'living fossil', because it had been discovered in rocks long before the actual living tree was found.) It has the distinction of growing its buds below the leafy twigs rather than below the leaf base. This is fairly academic as far as we are concerned as it grows imperceptibly and we've got a long wait to see the cinnamon-brown, deeply furrowed bark of what one day will be a fine columnar trunk. More homely by far, and certainly more suitable, are the medlar and bullace in our upper orchard. The former has so far produced several curious brown fruits which didn't stay to ripen, but it is supposed to make a good shade tree with flowers not unlike those of *Cydonia oblonga*, a common quince tree which we had in Greece, whose fragile flowers, fine as porcelain, would appear on gnarled twigs before the leaves. The delicious pungent fruit which I made into a coral-coloured purée, was hard as wood to cut up, and filled the house with an unforgettable aromatic scent.

Other fruit trees that we have planted yearly are apple, pear and plum, a fan-trained greengage, which has so far produced nothing but leaves, and about six damson trees. The latter act as security against the day when our old damsons, that have so ruggedly stood for generations around the house, and which are indigenous to this part of England, finally keel over. Damson fruit is some of the best. Its sharp taste upstages any milder plum or cherry jam, and damson purée, mixed with thick cream, is not only good to taste, but the streaked colour of dark purple and white standing in a green and pink bowl brings the sense of a summer garden into the house.

Two more walnuts have gone in since our first, which sadly lost its head in a storm and is now growing into a grotesque squat dwarf with not quite the grandeur we expect from a walnut tree. Birches and willows grow in gatherings about the place; the willows particularly leap into the air with tremendous verve, and in five years have reached thirty feet. The birches aren't so dashing, but they have already grown tall enough to give us an idea of shapes to come.

In the lower orchard along the boundary with our six-acre field we have a very ancient hedge of hawthorn and laburnum. A good deal of laburnum grows wild around here, even quite high up amongst hedges on the lower slopes of the hill. This old orchard hedge has grown immense; it is still possible to see where once, about fifty years ago, the trees were pleached by almost felling the hawthorns but not quite cutting right through, so that they could be bent sideways and threaded into the next one to form a stock-proof hedge. Now, however, having long since been left to grow untrimmed, the hawthorn and laburnum have reached over thirty feet. By constantly cutting them back in a certain shape, Michael has made a tunnel. Here, in June, we can walk under the massive hawthorn blossom and hear the murmurous haunt of bees, not flies, as they gorge themselves on honey. We would never have contrived or even thought of making a roughly-shaped

yellow and white tunnel like this, it just evolved, but it is the way so many of the best things have come about, almost when we were looking the other way.

There are just three more trees of our unending planting that I want to mention. One is a Whitebeam, *Sorbus aria* 'Lutescens', a really eye-catching specimen in spring, when its leaves are covered with a woolly mat of whitish fur. Involuntarily we are drawn towards it, imagining it has turned into a magnolia overnight. From a distance the leaves look just like huge unopened flowers. Later the silvery-grey appearance deepens to green, yet throughout the summer its magnificent shape and foliage make it a superb tree, with bright red berries for the birds in autumn.

The way a tree grows has so much to do with one's enjoyment of it. This is quite apart from its flowers or foliage; instead it's to do with its outline. Surely that is why a row of Lombardy poplars creates such serenity in a large landscape; or why a great swatch of flowers seen from under a rounded head of some tree, has far greater impact than if it didn't have this added dimension. Everyone knows the beauty of bare branches in the winter countryside; or the layering fresh greenness of beech boughs in spring. So in thinking of trees for the garden, it isn't only the catalogue description of height and autumn tints that matters, but also shape. Shape is a difficult ingredient to assess when making a choice for what must be limited to perhaps merely two or three trees, and yet one which is so crucial. So the form of one of our trees is entirely what influenced our decision. It was a Flowering Crab, *Malus floribunda*, which we had seen as a mature tree in someone's garden, where its disparate tangled growth of wide-spreading branches gave it an unusual outline in contrast to straight lines. We planted ours near the granary, where the angled outside stone steps are seen grey and worn from under this totally ungeometric tree. In spring it has astonishing crimson petals on the outside which, when open, reveal their white interiors. This malus is not an original, rare or exotic tree; it is nothing special and is to be found everywhere, but its character is one of distinction. And because I have planted the autumn-flowering white *clematis flammula* to grow into the branches, it has what appears to be a whole new blossoming late in the year, when spring seems to emerge all over again.

Finally, because I must leave the subject of trees, there is one that I think will be a failure. Not because it isn't a beauty, as we have seen in other gardens, but because once more we are defeated by the weather. It is a *Prunus subhirtella* 'Autumnalis' or Autumn Cherry, which I know is a gift in winter with its invaluable long-lasting white double flowers from November to March. However that is not how it works. For us at least. If, by January, it has at last managed a few buds, unless I cut down the tree and bring its whole head indoors, where buds will open with abandoned gratitude for the warmth, the first blast of frost burns up those frail little whitenesses into what looks like a lot of desiccated spiders. What is the point of such a tree standing high in winter just outside our windows?

Before moving on from trees, a quick word about yews – they are not as recalcitrant as their reputation for being slow growers suggests: the bad press they get should be taken with a pinch of salt. A hedge we have planted on two sides of a small sheltered stone garden is doing well. We are truly impressed with the way the yews have matured. Already they have created a four-foot hedge with shape and stability from eight-inch things we put in four years ago.

We shall go on planting trees, on into the future. We have the space and still have the energy, so they are a part of the garden which can be on-going in a way that increasing the number of flowers, roses and climbers would mean a long, serious think as to just what we can undertake. But trees are monuments. Once the decisions have been made, the roots spread out and the compost laid, then you need only stand back for sixty years. It has great charm that thought for gardens in the mind.

2 The Slow Infiltration of Water

Vicarious Gardening / The Picture Garden / The Benison of Saints / Water and Wildlife / The Kingfisher / The Infinite Stream / Our 'Temperate' Weather / Insouciant Disarray and the Perversity of Rareness / Errant Flowers / Childhood Scents / Michael's Scything / The Plant Collectors / Fanatical Priests / Extra-Floral Vision

Like an eel returning to the Sargasso Sea, my thoughts return over and over to certain pictures in a book, *Private Gardens of France*. They are of a garden in Provence where the photographs, clear and imaginative, show a garden of stones and greenness. Just that. Not a flower to be seen. In one there is a circular pool, very simple – a perfect O – with a narrow stone margin, nothing else, but reflected in its surface are two tall cypress trees. At least I take them to be cypress trees because of the shape, but as the picture is in black and white all that can be deduced are the straight trunks and sombre shade. I wonder whether the owner of the garden was influenced by the secret terrace of long ago, the *giardino segreto* – a hangover from the medieval cloister – designed by Michelozze for Cosimo de Medici at Fiesole? The calm serenity of this planting, where the owner with heroic restraint held back his hand from using flowers, is immediate and comprehensive. Beyond the trees is a misty view evoking the heat of the Midi. It is a landscape of cypress and olive overlaid with the quality of hushed midday, when everything except cicadas is in deep repose.

All the pictures of this garden save one – and there are twelve in total –

show the grace resulting from the mere use of stone and trees, terraces and vines, a mossy fountain and shrubs. The one exception shows flowers, terracotta pots of camellias, beside a flight of steps. This is a garden, yet it is lacking so many things we attribute to gardens. When we use the word we think first of flowers. How paltry it is to have only this one word – garden – to describe so many diverse creations. For instance there are wild ones and sanctuaries, grottoes, cloisters and medieval enclosures; there is the Renaissance plaisance, and there are front and back gardens. Our vision of paradise can be a five-by-five concrete yard facing north, but it is still a place to sit over a drink, where jasmine might grow and coloured pots clutter up a corner. A friend in Greece has his plot outside the back door where, under a vine, sitting on a rush chair, he looks down at his gaudy geraniums and a tin or two planted with chives. Beyond are old tyres, discarded plastic bags, a few undernourished chickens and last month's newspapers. That's his neighbour's yard. His own is self-contained and illusory; a place for sunlight and a pot of basil.

So this over-used word 'garden' has to do for all manner of places, and when I go back to those Provençal pictures of a garden to appreciate again the elegance of design and texture, I know well that here is a different sort of garden. If I had another gardening lifetime I would make a disciplined creation inspired by this one, depending only on stone, water, shadows and the use of leaves.

How frustrating to be limited to one garden at a time. There are so many temptations. I would like to dibble myself into diverse places of climate, outlook or continent and leave parts of me growing there to live a separate life. In this way it would be possible to have many options, including one something like this garden in Provence. People who move every ten or more years after having made a garden, are indomitable. How can they bear to pull up their tap-root and start again? What of endeavour? Of realization and assiduity? Do they return every few years to look over a wall to see how their *Caryopteris* × *clandonensis* or their *Elaeagnus angustifolia* is doing? Is starting a garden, the planning, designing and planting, the arrival in itself? Or are there altruistic gardeners who can get a vicarious pleasure from watching someone else take over what they have started?

When I meet someone beginning once again on their fifth or sixth garden I wilt with weariness at their perseverance. They radiate plucky resolution for what, after all, must be a mammoth chore. Cooking is a bit like that. The relentless doing it all over again. Each time you've got to peel the garlic

before getting on to the nice bits. On the other hand, with growing things, if you move from chalk to clay, from loam to sand, the true gardener is burning with eagerness to have a chance to grow those things banned from their previous place. Is that the incentive? To me my spirit flags even at the start. I just want to go round the world peeling off little cuttings to plant here and there, to say 'Grow!' and leave it at that. Years later, I would wander from one of my sylvan productions to another, each one absolutely reeking with success.

As I cannot make an abundance of gardens, it is back to that book and its pictures. Pictures in a book are a good substitute, and if you think of it objectively, it does save an awful lot of bother. I know, for instance, that I'll never make that absolutely dazzling herbaceous border shown in Marina Schinz's *Visions of Paradise*, or any of those wild flower patches photographed in Miriam Rothschild's garden from *The New Englishwoman's Garden* by Alvilde Lees-Milne and Rosemary Verey.

It's not just pictures though, words get to you, too. Writers like Jane Brown, Beth Chatto and Robin Lane Fox, Penelope Hobhouse, Christopher Lloyd and Rosemary Verey have a lot to answer for. They tantalize you with their good sense and encouragement. Their rousing words conjure up a fragrant and floral sequence unravelling through the years. Reading books about gardens is a potent pastime; books nourish a gardener's mind in the same way as manure nourishes plants. I feel I need this mental compost – not just for the hard information that some books are filled with, but for the imaginative writers who present me with variations and fire my resolve. Sitting in a chair I can let my mind be elated by all sorts of outlandish permutations.

Much as I'm a soft touch for garden books – using them in winter for my sort of vicarious gardening – I admit that there are some publications that are abysmal. They have appeared recently like a crop of deadly fungi. These are the gardening books where the text is negligible, not because the authors are not worth reading, but because the publishers have relegated their text to amputated columns interspersed with senseless pictures. The author mentions a cauliflower, so there has to be a photograph of one displayed on linen; if the word 'digging' crops up, the flow is broken with a photograph of a foot on a spade. Trowel, rake or hoe – they all need to be depicted. It's almost as if there are trigger words that set a publisher off, saying: 'Right. That must be illustrated.'

But why assume the reader is a half-wit? Do we really need to see a picture of a ball of string to comprehend the meaning of the words? The BBC is bad enough. If a car is mentioned in some feature on travel, we have to endure the noise of changing gears. Do words no longer have value? Do publishers take us for numskulls; for readers so goofy in the upper storey that we can't assimilate six lines without an illustration to elucidate the meaning?

As for garden photographers, how differently they see things. With what ease the camera seems to compose a picture of great beauty with its discriminating lens. The naked eye can't censor some ugly sight on the periphery of vision; the photographer takes the perfect shot and picks for us just what we need to see.

Then there is the enormous relief when the opposite happens. The satisfaction of looking at pictures of gardens which you are absolutely certain, instantly and instinctively, that you do not want. Everything you see is hideous. They are the pinnacle of brilliant undesirability. What a relief. Reading what some people do to their gardens is like struggling with another language; you can't imagine what they are getting at. It may have something to do with bedding plants, or perhaps not that, but rather the way bedding plants are used. And that is very elusive. Ordered blobs arranged in rows by heights and alternative colours so dispiriting that you can stand on the threshold of the garden and feel poorly. Yet in another garden the bedding plants can be collected and grouped to make the most dazzling, high summer clamour of colour by clever blending of blues and purples, crimsons and creams.

In India we've seen a most original effect from bedding plants. Petunias, in particular, looked wonderful in the imaginative way in which pots were grouped in concentric circles, on three low tiers. Shades of mulberry, heliotrope, ultramarine and amethyst were arranged, one circle within the other, into floral domes; some were carmine, madder, scarlet and clashing magenta, others were rings of topaz and lemon, highlighted by pots of pure white. Set amongst the deep shadows of citrus trees and patches of brilliant sunlight, these graduated mounds appeared as huge semi-precious ornaments. The flowers looked so passionate I almost expected their stalks to be hot. Why not, with all the sap coursing through their stems? An effusive-looking peony of the flaming whorl of *Lonicera* × *tellmanniana* exude amorous ebullience.

You don't have to think hard about flowers to like them. The scale of flower likeability is immense. You can like a daisy or feel fervent over an acidanthera. Sometimes, though, I wonder what I am looking at; I don't mean the sheer ignorance of plant names, but that a single flower, a clump or arrangement, can delight us instantly with sensual awareness of its rightness. But what is it exactly that makes that particular thing have such immediacy? It is hard to analyse. Gardens are made up of so many attributes: colours can be as potent as taste – the almost brittle fineness of the cups of a dark plum hellebore with its yellow stamens like silk filaments; the pale toffee light of a bleached thistle; the balance and complement from contrasted green textures – from shapes of crenellated leaves or from smooth arcs. Then there is scent – this added dimension to any garden. If in your wildest, most creative, moments you had invented a rose, would you have had the recklessness to add scent? So what are we appreciating in a garden? The details or the effect? The alchemy of gardens again – it's woe to anyone who wants something else in life.

This is why visiting gardens by book rather than by car can be an all-the-year indulgence. It is why winters are so imperative. And why I return again and again to that picture in the book of a garden in Provence, knowing that whatever my mood, whatever the season, its calming appearance is there, petrified within the pages on a shelf.

There can be no other occupation like gardening in which, if you were to creep up behind someone at their work, you would find them smiling. I know this as a fact. On a grey day in winter I found fresh, fat green buds of *Clematis armandii* about to break into small, scented white flowers. I put out my hand to touch them gently, and turning away knew I had an idiotic smile, doting and pathetic, on my face. I recognized that expression as one I've seen on others. We project love and felicity. Our tender cherishing goes out in almost tangible bouts as we stoop towards a plant, or reach to stroke a petal. The reverse does happen of course. In summer, when I discover a plant wilting to death because a mole has bulldozed its roots, the tension and bad feeling is a hundredfold. Then it's our cat tiptoeing silently across the lawn that brings back the imbecilic smile. Sitting on the ground to stroke her warm fur in the middle of frantic despair over irascible moles, once more benign serenity overcomes me and I think I won't offer the whole damn garden to my neighbour's goats after all.

It's at moments like these we could do with Saint Fiacre de Breuil. But why does he keep such a low profile? After all he is the patron saint of gardeners. Shouldn't we each have a small figure of him blessing our gardens from some secluded corner? A little homage to this gentle saint, who is usually shown either holding his spade or digging, might be just as efficacious as a blast of Phostrogen about the place. But what an enigmatic saint he is. Gardening books don't seem to mention him. Beyond the fact that his feast day comes at a rather drab time of the year, the thirtieth of August, I can't find out anything of his history. Yet there can't be any of us who might not be glad of a touch of his beneficence now and then.

Should Saint Fiacre fail there are others. Self-effacing saints of gardening lurk about the centuries. According to Dawn MacLeod in her marvellous book *Down To Earth Women*, the bluestocking eccentric Eleanour Sinclair Rohde, claimed that the earliest patron saint of gardeners was Phocas. She relates a noble story of how he allowed himself to be slain and buried in a grave he had already prepared for himself amongst his flowery plot. Representations of him are to be found outside the cathedral of Palermo and among the mosaics in St Mark's.

That was in the third century; Saint Fiacre, who extends his protection from gardeners to Paris cab-drivers, lived in the sixth century; Saint Maurilius, another beatific saint appears in a tapestry in Angers from about 1460 'with a halo and a violet-coloured robe, digging with a long-handled spade.' There also seems to be an even more elusive saint stalking the shadows of our gardens, Saint Sylvanus. With a name like that a small plaque to him hanging round a tree in the corner of the garden might do us a world of good.

Of all places in the garden perhaps it is our pond which brings about a recurring intensity of pleasure when we survey it. Water is compulsive; it draws each of us to gaze transfixed in a becalmed state which few other things induce so forcibly. When I first suggested making a pond in the fold-yard, Michael, always reluctant to use strident and intrusive machines if there is an alternative way, thought we might dig it by hand. We did try – but why is it we have such exalted views of our abilities? We are always being caught, thinking we can do things that we can't. Digging the pond was one of them, it turned out to be far too big a task. In the end we had the area dug by machine into a roughly oval shape, about thirty-five feet long, and to a depth of about three feet at its deepest. The puddled, clayey bottom keeps the water from seeping away completely and we top it up by pump from the brook.

Partly cobbled, full of half bricks and tiles, old horseshoes and rusting gate hinges; tufted over with indestructible grass, docks and nettles, the surround to the pond was too much to cultivate in one go. Instead I planted here and there, slowly ousting the weeds as I went along. Each time I dug a hole, in went compost and some better soil so that gradually, since March 1980 when we first dug it out, the whole yard has been dense with yellow, blue and white irises, mimuluses with their velvety open mouths, phlox and ferns, astilbes, marsh marigolds and legions of primulas including rosy-purple *P. sieboldii*, yellow *P. florindae* and small, dusty-leafed *P. frondosa*.

The whole of this lush verdancy is what we worked for and what we hoped for, but what were drawn to the pond from across the fields and through the air – blown by wind or brought by birds – were the hundreds of creatures that appropriated the pond as if by magic. Frogs were the first to find us. Surreptitiously their spawn appeared so that now it's hard to believe that each frog has to mature for three or four years before breeding. Every summer all over the garden we find their inch-long progeny in unlikely places as we clear under plants or move stones. Newts found their way here and often, when I'm sludging about in the water clearing the weed out handful by handful so as not to destroy a creature, I pick up one of these timid newts with a tummy the pattern of a tabby cat. By the end of July they've taken to the land to remain hidden in damp places throughout their winter hibernation. At other times I gather up in my hand from the pond the most voracious and evil-looking predators, which I put back into the water because one day these quite inappropriately named nymphs will turn into shimmering dragonflies and slender damsel flies that hover and dart so swiftly it is hard to follow their flight. Innumerable pond skaters, back swimmers and snails are forever on the move; the skaters skim over the water causing the surface to shimmer with continuous minute dimples.

Swallows and wagtails are constantly around. The grey wagtail in particular struts and wags about on the blanket weed which maddeningly appears every May, and like great soaking wodges of dishcloth, has to be dragged out each year. On midsummer mornings a solitary moorhen stealthily glides round, hiding in the margin leaves if one of us so much as moves at the window. We think it's a young bird each year by its brownish-grey colouring, but however hard we try to find out where it comes from or where it goes, we never do. Dippers which fly up and down the stream and sing in a continuous rippling sequence seldom come to the pond. Only in the early mornings will we occasionally see one meditating momentarily on a tiny island, built for wagtails, in the middle of the pond, before making its low, rapid flight back to the stream. There we often see them bobbing their stocky bodies up and down on a stone before diving in and spending what seems a hair-raising length of time under water. Sometimes we surprise a grey heron; it rises lethargically to disappear across the garden with lazy wing beats and nothing, thankfully, in its beak. For we do have fish, too.

But more dramatic than any other wild creature that arrives to charm or intrigue us is the kingfisher. Because the house stands high above the fold-yard, from the kitchen windows we can look down on to the pond. From here, when I'm involved in some mundane chore, boring and stultifying, I suddenly catch sight of this radiant bird like a ribbon of brilliant light. What a mercurial creature, appearing like a spectacular foreigner with plumage of pure cobalt blue and bright chestnut, far too flamboyant for our mild countryside. It has no sense of reticence but lets out an unmistakeable and piercing 'chee-kee' so that we always know it's arrived. Sometimes it alights on a *Rosa glauca* right under the window, so that I feel by stretching out my hand I could almost touch its iridescent back. At other times it hovers momentarily over the pond then plunges in headlong. Without pausing it comes flying up through the water into the upper air.

At other times the kingfisher perches on a slender forked branch pushed into the bank, remaining motionless during the afternoon. How clearly we can then see the faultless arrangement of feathers, the white flashes behind its eyes which almost join at the back, forming a peak for the lustrous flecks on its head. With wings folded only the bold line of the spine is turquoise, but once the bird is hovering, static in space, for a few brief seconds the whole dazzling span of colour is visible. After its plunge, with a fish in its beak, the kingfisher flies to the top rail of the yard gate where it thrashes its prey between guzzling gulps of impatient rapacity. Occasionally it drops the fish, then, like any commonplace sparrow it has to land on the ground to finish it off.

On favourable days it can down a couple of two-inch fishes as well as whole helpings of fat tadpoles already equipped with flailing legs. Other days are a failure. Its luck runs out, but not its patience. Twenty minutes can be spent on a swaying seed-head of a mullein, while it moves its head like an Indian dancer elongating and recessing its neck to keep its balance. Or on a day when the wind tears in great gusts, ruffling the surface of the

pond, the kingfisher crouches on the edge of a stone, neck retracted, watching the water. It fluffs out its chest which droops cosily over its claws which, oddly, do look a little 'pigeon-toed'.

We tend in this country to think of the kingfisher, *Alcedinidae*, as a bird of rivers, but in Greece we would always see it streaking low across the sea. The beautiful legend of Halcyone who threw herself into the sea after her husband was drowned, tells how they were changed into birds and built their nest in winter on the water, charming the waves to remain calm and serene for fourteen days.

Meandering at the bottom of the yard and on through the lower orchard for about two hundred yards or so is the stream. Along the banks are indigenous alders, handsome trees with shiny leaves, which lend the right height and lightness of foliage. Other natives, though, are not so welcome. Sycamores, with their greedy roots and impenetrable shade, seed far too freely. Neither have I lost my heart to ash trees; they are so grudging, not coming into leaf until the very end of May or early June and, even more mean-spirited, they are the first to lose their leaves. Sadly we had about twenty elms dying along the brook. Because we didn't want to do much to this part of the land other than clear it for replanting and encourage a natural wildlife habitat, we did need to get rid of the dead elms. Someone told us of a group of willing and well-organized conservation volunteers, who cope with this sort of work. One weekend in February they came, bringing their tree-felling equipment, and their food and sleeping bags for a night on the floor of the village hall. There were twenty-three of them, aged from seventeen to over sixty: one worked in the inland revenue, one was a long-distance truck driver, another worked in a factory and one huge chap was, by profession, a tree man yet still chose to spend his weekends in the open working for conservation. Some weekends they would be clearing ditches, others protecting sites of rare flowers or butterflies, or fencing against sheep on a hillside to defend a special habitat. After two days' work here they left us with every dead elm felled and piled into cut lengths, providing us with enough firewood for the next six years.

When we started to garden we had no experience of frost and of the devastation it can cause as it flows unimpeded from the upper orchard down to the stream, where it backs up to form a lake of wintry deaths. Buddleias and eucryphias, ceanothus and magnolias all perished as we planted them with ignorant optimism. It is the penalty of having a stream; we live at the bottom of things. Yet I know we are fortunate to have so much water in the garden. It brings the frost, yes, but it does add an additional breadth. Across the kitchen ceiling are dappled patterns of sunlight making continuous ripples reflected from the pond; from every room in the house in summer the brook is audible. But I feel it is with the stream we could do so much more.

Experienced, devoted and energetic gardeners would already have made fabulous bog gardens with gargantuan-leafed gunnera, or more reasonably sized and drably named skunk cabbages, ferny-leafed astilbes with their muted subtle tones, spires of filipendulas and so many others. But our thoughts have been turned in other directions. Only now am I slowly moving towards the stream by planting drumhead primulas and flag irises, divided from multiplying crowds in the yard. Clump by clump I'm adding handsome rodgersias and ligularias – plants grateful for the shade, the damp and the lush jungly undergrowth that overhangs the stream in places.

On the north-facing side of the stream things need little encouragement. Ferns and the shiny evergreen hart's tongue, *Phyllitis scolopendrium*, with strap-shaped leaves appears at the shady base of walls as well as along the bank and seems to get greener as winter progresses. That blessed pulmonaria, *P. officinalis*, with spotted leaves and pink and mauve flowers spreads avidly in the grass.

Stream constructions also hold an endless fascination. One year we built a low weir to hold back the water just long enough to form a smooth glassy area of calm and reflections. My extravagant plans for a barrage between the banks, imagining deep pools and a hurtling sound of tremendous water weight falling from level to level, are somewhat tempered by a dreary modicum of sense and the even more prosaic fact that the brook has a habit of gathering great shoals of silt on corners, as it winds on down through the land. There is one point at which I've built a dam of boulders, more for its gushing sound than for its efficiency in holding back the tide. Every year on some warm day in early summer I have to remake it. What a delightful job, plunging about in water so noisy it cuts off any outside world of telephone or human voice; groping under water for the stones and rocks that are annually tumbled by the winter turbulence. The whole edifice must be rebuilt each year so that we can enjoy the continuous sound from its small waterfall.

Another construction is the bridge. Originally I'd drawn a modest stone hump-backed affair which would sit so well between the banks, and would grow mossy and integrated to look as though it had been there since the cottage was built hundreds of years ago. But Michael needed a practical bridge, one over which he could drive his garden tractor, so it was made of longitudinal beams with criss-crossed railings of slender larch on either side; it leads right into a hexagonal summer house with seats around the walls. In summer we keep a table here so that we can have tea on rare, scorching days, peering out at the garden from under the rampaging rose, 'Rambling Rector', which covers the roof with billows of white flowers.

In shallow places across the stream we have stepping stones. If you are quiet and look towards the pools you can see small brown trout dart for cover. Like so much about gardening, there can never be an end. The possibilities with the stream are infinite; we can go on constructing and planting, for water is a powerful incentive. But whenever I feel a reckless bout of wild experimentation coming on, I do remember that Irish gardener William Robinson's acerbic comment in *The English Flower Garden*: 'People are so much led by showy descriptions in catalogues, and also by their own love for ugly things that we often see misuse by the waterside of variegated shrubs – a yellow Elder, the purple Beech, and even down to the very margin of the water with variegated shrubs, absolutely the worst kind of vegetation which could be chosen for such a place.'

Sometimes I wonder whether gardeners like gardening. Really. I don't mean my sort, but the real gardeners, the dedicated ones. The seedsman, the propagator, the plantsman or -woman. Do they like it or is there somewhere a kind of perverted satisfaction that comes from doing so much that is hateful? Because surely, if you are truthful, a great number of gardening jobs are pure slog. Many of them are carried out when the days are bitter, and there is the endless repetition of thinning out, or staking, or coping with greenfly or checking ties against wind damage and looking for somewhere to put your feet. What is likeable about it? Surely it is masochistic perversity to go through pure, tedious hell so that coming out the other side you can say, 'Ah. That's over', 'Look at that!', 'What a day I've had', etc. And woven through it is the sheer, unfunny business of having left the string, the secateurs, the spade on the other side of the garden. Of either hoping you can do a job without taking all your stuff with you, or cursing yourself as you walk back for the sixth time to fetch a trowel or a cane, or a container into which to put the unexpected stones that you've dug up, or the rose twigs too thorny to be put with the weeds committed to the compost.

Are there really people out there who can say, with hands on their hearts, they enjoy it? Be honest. I know one gardener at least who is. She is a brilliant gardener, but meet her in November and she admits to being glad not to garden. Like me she doesn't want to know; and we drink together happily considering the next few months when we are let off the hook.

Spring is a time for being manipulated. Everything is on the move as the days get longer and buds glisten with the lustre of renewal. Yet with spring there also comes apprehension: those freak warm days that hoax the plants; that familiar dread when a few hours of sunshine bring everything on too soon and even while you are admiring, and almost complacent, you know all is doomed. There will be frost tomorrow, or the next day, or worse still next week, when even more vitality has overtaken the *Hydrangea villosa* and its leaves are growing daily by centimetres. Go back, go back, do! This isn't May, this is still March. The mild weather is having you on. Go back, lie dormant, and please, please be patient!

Yet the gardeners whose reason tells them this, are at the same time going mad with gratification that blossom is appearing on the *Viburnum* × *burkwoodii*. And then wham, bang! Overnight the frost comes, the weather behaves seasonally, all the mauve and blue crocuses, the whites and the yellows have lain down into soggy pulp and we wonder at our idiotic foolishness in having spent hours and hours last autumn planting them. Yes, spring can be the devil, not only fooling the crocuses, but us too, when momentarily we forget to look coolly at the calendar.

I don't mind being at war with greenfly or black spot, slugs or mould, and long ago we accepted life with rabbits and moles, but I am not resigned to the weather. That enemy that we can never vanquish, that adversary which lays its menace over every garden endeavour. It fights unfairly, for every year the weather is pronounced 'exceptional'.

Our temperate, seasonal climate that produces such beautiful English gardens. Temperate? What rot. It's war; a continuing war of droughts, ice and boreal winds. Long ago the sane gardener acknowledged that the exceptional is normal. There hasn't been a year when there aren't at least two exceptional periods. This is so patent, so obvious to anyone who has tried gardening for a few years, that any book which tells you in this or that month something should be done, is having you on. For you can be certain that if you follow the advice and cut back your buddleia in March, according to the book, those darling little lively leaves, stimulated by your pruning, will somehow one night change into friable fragments that turn to dust in

your fingers. Oh yes, but ... it was exceptional. Exceptional? Read it for NORMAL. That is what our climate is. That is the truth. Do you remember the five weeks or so of pure ice? 1986, I think. In some places the ice went down to a foot and stayed there. Or 1987 with its bitter winds coming late and, naturally, exceptionally so, followed by the catastrophic hurricane in October? Or those exceptional droughts when there was panic at the Water Board – or frost in May? No, no, you've got it all wrong. These are all normal. It is us who are abnormal. We won't face up to the fact that we should be ducking and dodging, conniving and being canny to outwit the weather. We should all be cynics. We should never believe those mild days when the wind blows like a southern zephyr and we smile idiotically because spring has arrived. Think of April 1987 when the weather leapt into the seventies, and our spirits leapt, too. Cotton clothes and meals in the garden, the blossom and the bees, the buds and serious watering. If only we'd kept cool and sceptical, if only we'd known the rug was spread temporarily. And has it ever been otherwise? I haven't gardened long enough to know the answer. During the last eight years of making a garden, I only know that we have staggered from one exceptional bout to another, doing our best to coerce and control plants in this endless game of bamboozle the gardener.

But hold on just for a moment and suppose – imagine a climate in which after winter things did grow gently vernal; and sowing, planting and dividing did occur in sequence; then this indeed would be exceptional. Meanwhile I write in a *normal* spring surrounded by devastation: hebes, rock roses, lavender and thymes are burnt brown by inclemency. The ravages of hostility are all around us.

Why do experts go on so about lawns? Why all that time over the air or in print on the minutiae of worm casts and moss? What is the spell and fascination of spending such a disproportionate amount of time on this one gardening subject when so many others have more allure? Personally I equate making lawns with washing-up. You needn't be told twice. And if we are doomed to hear these relentless turgid explanations on how to grow a lawn, couldn't they please put in a whispered plea for negligence? A mild suggestion, a subliminal leak that part of a lawn could be given over to speedwell, that cerulean creeper which thrives in spring adding bloom across grass. Bloom on fruit is a state of innocence, why not on lawns?

In its way chaos is as powerful a component of gardens as topiary and the static authority of dark green angles and planes. Dishevelment plays as important a part as the strong design of predominating stone terraces laid in level spans. Contrast is fundamental. Sunlight and gloom, control and bedlam, the yin-yang complement is almost an intuitive ingredient that we recognize instantly. So what about a little insouciance about the place? Random seeding can sometimes be a godsend. What gardener doesn't make a mental genuflection on discovering a self-sown group of violas by the doorstep, or on finding a spire of deep blue Jacob's ladder under the blackish-crimson blooms of rose 'William Lobb'? Shouldn't we realign our sights and acknowledge the splendour of campion? Do we need to be so churlish towards dog's mercury? Arbitrary seeding is something to be desired, almost to be cultivated. Can't there be one part kept aside where wilderness can become a gardener's eldorado, where like dandelions gone to seed we can lose our heads and allow wild angelica to gain a foothold?

A friend once gave me a tiny magnifying glass so small that when its two lenses are shut away it measures less than an inch and a half. The filaments on a seed head of a *Clematis orientalis* appear like the silky tail of the silvery gibbon; the heart of a pelargonium is streaked so dramatically with crimson seepage it becomes anatomical; the hairs on a calyx of a 'Comte de Chambord' rose-bud are as precisely incised as a filigree brooch. Once seen through the lens, the complexity of flowers, their fine tracery, fragile fretwork and delicate colouring make it impossible to dismiss some weeds as contemptible; for under the microscope, we are taken into a phenomenal dimension. Yet it is hard to know where to designate the acceptable and unacceptable wild flowers in the garden. Dandelions, buttercups or celandines, each with their strong yellows and handsome flowers, could surely be tolerated if only they didn't have the rapacious habit of taking over? They infiltrate everywhere, choking and overpowering more tentative plants with their prodigality. And yet how odd it is. For what one country cultivates with diligence and passion, can become an undesirable intruder elsewhere. The perversity of rareness; the incentive to grow something that doesn't take kindly to our soil or climate; the very fact that seeds or plants are listed as being difficult, tender, elusive or rare in this country, makes some gardeners covetous with ambition. The challenge is irresistible.

In northern Thailand we have a friend who swoons with love for his few washed-out sweetpeas – a reminder of the gentler shores of Albion – anaemic and ailing amongst the frangipane trees with their lingum stems which open into heavily scented waxy-white flowers. Elsewhere in Thailand, in a town one hundred miles north of Bangkok, where we lived for three years, our landlord, who was certainly a dedicated gardener, was so thrilled to produce a bunch of grapes hanging on the vine that he filmed it with a cine-camera. And this in a land of mangoes, pineapples, papayas and durian.

The 'squaw carpet' ceanothus, *C. prostratus*, with its fishbone patterns of growth, creeps in rumpled masses on the hills of San Francisco where it is unregarded by the passers-by; hedges of fuchsias in County Cork had me longing hopelessly just to get one of those plants to make it through the winter here. As for the wild flowers that proliferate around the Mediterranean, here we desperately cherish a few of those sternbergias or a handful of cyclamens and watch them throw in the towel year after year. They make me wonder why we never took our Thai friends a bagful of seeds of rosebay. Compared with poinsettias and hibiscus how recherché it would have been.

In spring when the dippers are at their most melodious, singing along the stream while they bob from one stone to another, this is the time for the wild flowers on the daffodil bank – that part of the garden which rises steeply up the other side of the brook. Every year things appear differently. Ladies' smock is the most unpredictable. Feckless and impossible to coerce, one year it's in the long grass of the lower orchard amongst the tulips, another it struggles to be seen amongst all the things around the pond; last year, perversely, it was on this bank. Wood sorrels, the original Irish shamrock that close their leaves at night, stubbornly refuse to spread elsewhere but cling precariously to life in the fold of a fallen tree where creeping ground ivy, with its self-effacing blue flowers, sidles along the ground in shady places. In this area too are evil-smelling plants such as cuckoo pint, whose brilliant red berries are loved by birds, and the invasive, handsome butter-bur, thrusting up foot-long stems with pinkish flowers before its canopy of huge coarse leaves prevents pollination; pull this out only once to know the strong odour that lies at its roots. You need to brace yourself – the rank smell percolates behind your throat and eyes.

Bordering the lane in the hedgerow are wild roses with wide, straying branches covered in thorns, and white bryony, whose long tendrils sinuously twine along the hedge amongst the other seedlings that have settled here for years. In the yard, besides the things I have deliberately planted round the pond, are the vagrants, charming and admired, from herb robert which looks particularly good when its filigree foliage and red stems flatten its circumference against the grey stone wall where it seeds in crevices, to one secretive common spotted orchid. I particularly like prostrate creeping jenny which trickles its yellow flowers across cobbles and under irises, and also the discreet flowers rummaging about the place of scarlet pimpernel, which closes its petals before rain. As though carelessly crumbled about the garden are small fistfuls of violets – rogue plants with an instinct for the right place. And I know I should be ruthless at the first sign of caper spurge with its architectural presence and distinctive bearing but it grows, and it goes on growing, and it usually seeds itself in inappropriate positions. By the time it's reached more than three feet I know that once again I've unwisely succumbed to its bluish-green leaves and small nests of ripening capers.

There are so many more – like yellow archangel and fragrant meadow-sweet – each welcomed and loved; but I spew out hatred and everlasting perdition to hemlock. I know in hedgerows it's delightful, I appreciate its wayside umbels so divinely conceived, but finding it in February already verdurous under my most precious *Rosa sancta* makes me hopping mad.

Nice little things appear in our old stone walls: small, rusty-black ferns and common polypody, with brownish scales on the underside of its leaves. They have a way of choosing places to grow which are entirely right; unlike the caper spurge which seems so obtuse, these ferns present themselves in artistic little fringes along the wall where, in a hot summer, they dwindle to dust and I lament their passing, only to find them back again in no time after a drop of rain. There is one wild flower, the cowslip, that I long to see growing footloose, as it does bordering Austrian meadows in spring. We do have them; a few here and there on the banks, but not in teeming masses, not at large in the way I wanted, for jogging childhood memories.

Scents of childhood are often indefinable but their potency is instant. Children brought up amongst privet go into raptures of recollected infancy when they come across it in later years. It is an especially unique scent and, for us who don't have this association, it is hard to understand how fervour can be aroused by such a dim little leaf. Cowslips, wallflowers and primroses can be powerful reminders of childhood. The very fact of having such a damp climate as ours, in which the sun is seldom fiercely hot, makes the scents last longer. Fresh or fruity fragrances linger with more vigour than the scents from those exotics on the equator. The aromatic smells from the turpentine of pine needles or from the thymol of thymes have a pungency at midday which meets you across the Mediterranean before you get sight of landfall – but damp primroses upstage the lot. Their scent endures from dawn until twilight.

Stephen Lacey in his lovely article on *Scent Gardening* in an issue of that inspiring quarterly gardening journal, *Hortus*, writes about the effect of scents that '... can either be direct or immediate, drowning our senses in a surge of sugary vapour or they can be subtle and delayed, slowly wafting into consciousness, stirring our emotions and colouring our thoughts.' They certainly can. Surely no one can resist the heady evocation of herbs which can calm the pulse rate or awaken thoughts of eating. Amongst so many herbs, with their differing savours, is one that instantly takes me back to Thai markets; to those dim alleyways where the huge rosy-coloured paper umbrellas shaded the produce; where after leaving the fetid smell of fermenting fish paste and passing amongst the vegetable stalls, were the bundles of herbs tied with a grass thong or a strip of banana leaf – and one of the most overpowering was coriander. Still now when I smell it the piquancy of those markets is instantly recreated. Admittedly when I breathe deeply into a bunch of coriander it does have a slight whiff of horse's urine, however no one would be put off the smell of basil just because of Isabella's head.

Michael isn't an undefined presence about the garden, though it may seem like that having appeared briefly by name in some general comment. It is not like that at all. Michael holds this place together. The cohesion of one part of the garden to another is united by the fabric of his grass design. What happens in between are bits of embroidery, good or not so good according to the season, dexterity and luck. But grass, in its wide stretches, is the textile at which he works with his mowers and his scythe.

Scything is the occupation that gives him the greatest pleasure. On a still summer morning when the dew has barely evaporated he is out there making slow rhythmic sweeps that lay the grass in copious swaths. For Michael it is the most satisfying and the most soothing of any outside work. The whoosh of the blade, the folding of the cut stems and the constant sharpening of the blade with a whetstone, are each part of this traditional country labour. Throughout the summer his scythe hangs in the crook of an old pear tree in the orchard. No doubt generations of countrymen have laid their scythes there safe from the reach of children and animals.

Michael's grass-cutting design is a vital element to a garden such as this where there is little formality, no strong lines or defined boundaries. By carefully planning to allow wild flowers to seed in certain areas; by planting fritillaries and early species tulips as well as the later lily-flowering ones, and by leaving the tall grasses of fescue, Meadow Barley and Fine Bent, by late summer there is a pinkish bloom from their downy heads. Through this he cuts winding paths no wider than his mower, leading to seats. In the upper orchard the bench is under an ancient apple tree with a view looking down towards the house; in the lower orchard a path leads to a seat amongst a shrubby growth of filbert hazel bushes. Like so much about our garden, Michael's architectural grass-cutting evolved. He never planned it this way, but each year he goes on perfecting his design.

Early on, about April, he lays out the first paths, the ones that he'll keep short all through the summer. By June, when it's safe to cut down the dying daffodil leaves, he begins to mow the grass to a higher length which immediately adds a textural contrast to the overall effect. Not until August, after the seeds have ripened, are the two main meadows of tall grasses finally cut. The result, for a week or so, is unattractive, where patches of yellowish-brown destroy the harmonious greenness, but not for long.

Grass is a chore. People often ask us how we maintain such an informal effect. The answer is a paradox. Untidiness requires work. Beds with defined edges, neat and trim, make it easy to keep things under control. Years ago when Michael first started he realized he'd need a good machine, one to sit on not one to push up and down our undulating land. He bought a 'Wheelhorse', an invaluable contraption that cuts at different levels, scoops up the cut grass and has a small trailer for carting the hay or for moving stone for our endless productions, and for transporting mounds of compost when we mulch everything in spring. Then he has a very sturdy mower for the intricate bits, which is also a brute of a gadget. It is vicious, ear-splitting and maddeningly useful. I loathe it. When he's using it, it's as destructive to the atmosphere as it would be to water the garden with acid. As it is, its lethal little flail may garrotte a flower here and there. Yet the alternative to this devilish device is a small hand sickle. We both use one. Because so many shrubs are planted into grass and not into well-weeded flower-beds, it does mean endlessly having to cut the coarse grass to prevent it from sucking up the moisture from the roses. But to use a sickle for trimming the banks down to the stream or for keeping grass away from around young trees would be a job for Sisyphus. Michael's genocidal gadget destroys with undiscriminating professionalism. I know its value but I wish it would wear a tea-cosy over its works.

Both of us enjoy working with stone; Michael lays the paving, making the flat bits, and I make the walls. The system works well. We made a sunken garden in this way; we made the terrace and low wall outside the granary together. In our hexagonal summer house across the bridge, Michael laid a floor of beautiful, old rosy-coloured bricks radiating out from a central orb into six wedges that reflect the six sections of the roof. He also laid brick paths crossing each other in the herb garden. This was one small garden intended to be formal in design, but years ago it got out of hand; fennel, lovage, angelica and borage took over in avaricious haste. The effect of tousled abundance is so pretty we are very half-hearted about bringing it under control. Anyway every winter we can enjoy the subdued rightness of Michael's design even if it is submerged in greenery by summer.

Like so many couples of which both partners garden, it seems nothing was planned as to which of us should do what. Fortunately we don't enjoy doing the same things about the place, though there are plenty of jobs which we agree we both loathe. Tying-up against the walls is one, spraying is another. We don't use weed killers after having tried it our first year. Death was too indiscriminate and the smell appalling. Now we use organic products or soap-and-water sprays, or we just accept that greenfly, black spot and mould are a part of the rich tapestry of a garden.

Michael looks after the trees. He makes the compost, hates weeding, loves pruning and making pergolas, hates wrapping things against winter frost, loves the instant effect of mowing and doesn't go squelching about in the pond. What we do share, and nearly always agree on, are what plants we'll buy. The enjoyment of that is endless. It is spontaneously combusted by a few pictures. We only have to look in winter at some book of magnificent gardens to become ambitious, inspired and besotted. How effortless. It just needs writing down and we feel it's done. Only months later, when we look at a strange name under a heading 'Must Have', does the dithering begin. What was that plant, why did we choose it, and far more crucial – where was it to go? That garden enticement of going headlong for a plant before thinking where it is to be planted, is folly and madness – but what a compelling winter pastime.

It is not one's partner's idleness or unfair apportionment of labour that may cut a marriage asunder, but it is wheelbarrows. Those lovely little three-legged objects which are the backbone of garden transport. No matter how many we have, and by now we have acquired four, they will be full, in the wrong place or mysteriously mislaid. They are the thin skin of garden compatibility, the Achilles' heel. Abuse, accusation and devious manipulation hang around wheelbarrows like bad vibes round the Moonies. Has any gardening couple ever owned enough wheelbarrows? I'd love to know. And if they have, what is the number?

Wheelbarrows may be a dicey subject between us, but there are the shared moments, the good moments, the recurring ones that appear in the form of small surprises to grandiose achievements. They may come when a new plant blooms for the first time. Our first erythroniums bowled us over. They're so unobtrusive, with brittle stems carrying fragile flowers, that to see them we needed to crouch beside them and turn up their yellow faces of curling petals. In contrast to this small triumph, there is our pond which never stops surprising us with its versatility. Our shared pleasure in its seasonal variation is a complacent certainty.

Mostly I don't know the names of my plants. Except for roses, nothing sticks in my head, and a long-suffering gardening neighbour with infinite knowledge and patience answers my sporadic and often repetitive queries as to what this or that is. So it's always a surprising pleasure when a visitor to the garden says, 'Oh, that's an especially good *Galtonia candicans* you've got growing there.' 'Is it, and have I? Please show me where.' Of course this is shaming and lazy of me not to learn the names of what I grow, and though I love gardens and looking at them, I know that never in my wildest enterprises around this place will I ever be a proper gardener. Improper I shall remain, and I have a feeling there may be many like me, who passionately love all sorts of aspects of gardening without committing their whole attention and concentration to that one particular subject.

Not so the plant collectors. These people I do revere. So many of our ordinary plants about the place, taken for granted – if ever thought about at all – as probably indigenous, have been gathered into this congested island full of its nine million gardeners by intrepid, single-minded fanatics who did care desperately what their plants were called. They needed to be labelled, especially as many plants were named after the explorer himself.

So walking round our garden any day, we can make a nod of gratitude to those pioneering zealots. I stand in autumn looking at the sorbus 'Joseph Rock', with its hues of coppery-purple and clusters of yellow fruit turning to amber, and think of the American plant hunter in China, about whom there is a book with the intriguing subtitle, *The Turbulent Career of Joseph Rock*. Surely this docile little tree didn't cause the turbulence? Two other Americans, John Bartram and his son William, were eighteenth-century

collectors who sent seeds and plants to England, amongst them *Hydrangea quercifolia*, with its distinctive leaves and floppy habit. Charles Maries in the nineteenth century, who finished his life managing the gardens of the Maharajah of Gwalior, introduced plants from China and Japan, including *Daphne genkwa*, with its violet flowers, and the fir tree *Abies mariesii*. Also the deliciously scented winter-flowering *Hamamelis mollis*, which has given up the ghost with us on two occasions so we've almost capitulated.

Stephen Haw, a Chinese scholar who leads botanical tours to China, wrote an article in *Hortus* entitled 'Memories of the Plant Hunters', in which he stated that: 'It was in western China and neighbouring lands that the greatest harvest of plants was gathered. The amazing richness and variety of the Chinese flora has to be seen to be believed; a single mountain in Sichuan, for example, supports more plant species than occur in the whole of the British Isles.'

Sir Joseph Banks (1743–1820), the great English naturalist, travelled round the world with Cook. A pliable climbing rose from China, *R. banksiae*, which has small white flowers, is named after Lady Banks. We tried to grow the rose here, having successfully had it billowing amongst the branches of an olive tree in Greece, but we must be content that the 'Yellow Banksia', *R. banksiae* 'Lutea', does just hang on to life on a west wall. Perhaps one hot summer it will achieve more than six blooms.

John Jeffrey, another Englishman, collected *Pinus jeffreyi*, a tall impressive tree with glaucous young shoots; and the famous Scotsman, David Douglas, whose travels in North America read like an adventure yarn, gave us not only the 'Noble Fir', but so many plants which are common in our gardens: *Ammelanchier florida*, *Rubus spectabilis*, the long-flowering *Clarkia pulchella*, *Mimulus moschatus*, and *Garrya elliptica*, which Michael and I are still waiting to see produce catkins drooping like grey-green stalactites. The Scottish plant collectors seem to have the edge on most other nationalities; there were many of them. Robert Fortune (1812–1880) was one who set off for three years in China after the Opium Wars, and endured unbelievable disasters from assault and robbery to typhoon and fevers for the sake of white wisteria amongst others. He also brought back to Britain, amongst his superb collection, winter-flowering honeysuckles, weigela and forsythia.

Long, long ago, in 1575, *Drimys winteri*, a small tree of tender susceptibilities but smelling so lovely with its scented ivory flowers, was brought back by Captain Winter, one of Drake's men. Surrey gardens have never looked back since Sir Joseph Hooker (1817–1911) collected rhododendrons from the Himalayas. The demure and small *Magnolia stellata*, which I thought ought to grow here in our garden, but which decided it wouldn't, was brought by J.G. Veitch (1839–1870), one of the great family of nurserymen who sent the two Lobbs, William and Thomas, travelling through Brazil, Burma and the East Indies to gather specimens for his famous collection.

Other renowned explorers were the Tradescant father and son who introduced *Tradescantia virginiana*, 'Spider Wort', and the Virginia creeper, *Parthenocissus quinquefolia*, which turns to such a dashing shade of brilliant coral on our barn in autumn. To John Tradescant the younger we are indebted for one of our most beloved trees that thrives in an informal setting and grows ten inches a year, the tulip tree, *Liriodendron tulipifera*, with its curious-shaped leaves that turn to celandine yellow in autumn; and another tree which conversely is reluctant to grow for us, the swamp cypress, *Taxodium distichum*, a most imposing tree with a reddish-brown bark.

There were a number of priests who were plant collectors, too. One was Pierre le Chéron d'Incarville (1706–1757), a Jesuit botanist who travelled in China and discovered *Syringa villosa, Thuja orientalis, Lilium tenuifolium* and *Deutzia parviflora* amongst many others, while he delighted the Chinese Emperor Ch'ien Lung by bringing him back from Europe the sensitive plant *Mimosa pudica*. Named after Pierre le Chéron is the volatile pink, trumpet-flowered *Incarvillea mairei* var. *gandiflora*, which is too delicate for our part of England.

Priests and botanizing seem to weld harmoniously. I remember a Jesuit who led us one hot season in April through the jungles of Laos. Perhaps it is their spare forms that give them the stamina and agility needed to pursue so doggedly their botanical lodestars. Certainly our priest, accompanied rather incongruously by his cocker spaniel, was tall and fine as bamboo; his black habit kept catching on thorns, his visionary eye was fixed on the congested foliage overhead while our eyes were kept on the matted undergrowth that impeded every step. Like other botanists we've met, his eyes saw things almost before they became apparent. Pointing there, high up, barely visible were the starry flowers of orchids. Like a passing shadow he was up the tree and then, in an instant, down again, with his cummerbund loosely holding the fragile gold flowers. We were five hours on that mountain. Long before the priest had satiated his craving for orchids, Michael and I and the dog had collapsed senseless in the heat, our faces plagued by small wild bees.

Perhaps the plant collector most often referred to is Ernest Henry Wilson (1876–1930), that indefatigable enthusiast who travelled and endured such appalling hardship in China. Amongst his many plants that flourish in our English gardens are *Lilium regale*, with its obligingly hardy habit and yellow-centred, funnel flowers of intense fragrance. Thanks to him, too, we have the many evergreen honeysuckle hedges of *Lonicera nitida*, with their unimposing flowers. Beside our pond grow graceful, long-lasting *Primula bulleyana*, named after A.K. Bulley who started the seed firm of Bees Limited, and who sent George Forrest, another famous collector, off to China in 1904. *Primula forrestii* was brought back by him. Another of Bulley's recruits was Kingdon Ward, who, in the twenties in Tibet, introduced *Meconopsis betonicifolia*, a flower of unsurpassed beauty, of such indissoluble blue, I long to see it growing in spectacular abundance and yet it seems so tricky to establish in some gardens. Kingdon Ward also gave us the giant cowslip, *Primula florindae*, a pretty plant with stout stems and bell-shaped flowers.

There were so many more of these explorers. Standing in the garden looking perhaps at one of George Forrest's *Primula vialii* from some distant part of Szechwan or Yunnan, now growing in alien surroundings, it is impossible not to feel a quickening of interest. Who were they? What were they like? How did they travel? Did they have pouches of soggy verdure strapped to their waists or hanging across their backs as they plodded, single-minded and hawk-eyed across the tundra; or made precarious climbs disregarding common sense and safety for the sake of some unnamed plant? Did they have a sense of humour, or ever get pleasure from a sybaritic life of warmth, food and friends? Or, at home, even in the midst of cosiness and antimacassars was there always a steely centre, a hard and relentless spirit willing them to pull on their woolly hats and be off in the teeth of a gale to Spitsbergen? We take them all for granted now, but I will stop to make obeisance now and then to Wilson and Forrest, Bartram and Bulley. The ghosts of these intrepid men have turned into flowers.

Collecting plants has long since become a form of outlawed vandalism. We can look, discover and record but not collect wild plants. Anyone nowadays caught wearing a 'vasculum' slung around their person would be hounded from the land as an immoral miscreant, for this object was the plant collectors' container for the specimens they gathered. And when I first heard of the 'Wardian Case' (the earlier version of the collector's box), I imagined it to be some kind of political scandal, some sleazy litigation going through the courts. In fact it was the brilliant discovery by Nathaniel Bagshaw Ward in the nineteenth century, that a sealed glass case created just the right environment in which plants and seeds could survive by re-cycling their own transpired moisture. This meant that living plants could be brought from faraway places to add not only to the great collections of dried and pressed flowers preserved in herbaria, such as the ones at Kew or in the British Museum (Natural History) botany library, but also to gardens.

These days anyone seen with their head down pacing amongst the sedges and uplands, or bent double in some flowery mead would perhaps be carrying a camera or sketch pad. Plant gathering may be ostracized, but love and pursuit are as ardent as ever.

One such person is Jo Dunn, an amateur enthusiast who has observed and recorded wild flowers most of her life. Talk to her and you are drawn into another world. Just as my tiny lens given to me by a friend has shown me a different perception, so, too, talking to Jo has made me realize that there is a completely alternative way of going about the countryside. Like the Jesuit in the jungles of Laos she sees things of such miniscule variation you feel her cornea, iris and pupil must be made of miraculous stuff. I don't think I could ever see what she sees. For one thing I'd need to carry a memory of what I'd seen before – an hour ago, last week or last year – otherwise how would I know if this particular flower were any different from what I'd seen in hundreds of places? One of Jo's discoveries was downy woundwort, *Stachys germanica L.*, growing in a new site in Oxfordshire. As 'Its known sites have dwindled to less than half-a-dozen' it was a thrilling discovery, and duly her account was published in *Watsonia*, the Journal of the Botanical Society of the British Isles.

Such love and passionate pursuit of plants must still originate from the same source as that of the plant collectors; when Jo travels in winter to Dorset to look at the Spring Snowflake, *Leucojum vernum*, for its rarity and beauty, over her shoulder may be the shade of Pierre Belon who, in 1546, set out for Crete on a spirited quest for the white oleander.

3 Stone, Walls and Climbers

The Generosity of Gardeners / The Penstemon Ladies / Garden Visits / Next Time Think Bigger and Bolder / Verticals and Design / A Gentle Plea for Chaos / Enclosing a View / Gardens in America / Wall Climbers / Precious Stones / On the Terraces / Unexpected Bonuses

'At last! I found you through the Conservative Society.' 'That's odd,' I said, 'because I don't belong. Never have,' which just goes to show how parochial we are although we live in such a large county. This lady, Mrs Porges, who had so deviously unearthed us, had a large garden of huge walls, topiary, ancient yews and a lake. We had been to visit it on an open day, and although she was almost eighty and suffered badly from arthritis, she wouldn't accept sympathy. Her acerbic rejoinders checked any compassionate noises we were about to make. Walking upright, with the help of an elegant stick, she showed us round, explaining that she still did a fair amount of the work herself. Apart from her swans, a superb vegetable garden and a bank of roses, what I remember most was my first sight of a *Hydrangea villosa*; such a large shrub with exquisite mauvey-blue corymbs – flat-topped clusters of tiny flowers – I have hankered for one ever since.

That visit on the open day had taken place three weeks before the telephone call; now, out of pure generosity and tenacity, she had taken the trouble to trace us because she remembered we were starting a garden. Ann Porges typifies gardeners. They are magnanimous and benevolent and quite unlike any other species. Travellers keep their sequestered finds secret, cooks are cagey about recipes, and as for archaeologists, they seem paranoid about revealing the lip of a jug or a sliver of scraper, but gardeners – they are something else. Ann was only the first of a whole unravelling of encouragement and support which came mainly from women.

Because I'm a flighty gardener I cannot remember any names or where I've planted things, so from the beginning I've kept a book in which I write down the date and place of planting, and the origin of each plant. Alongside this information are photographs of all the 'befores' and 'afters', in the hope that as the garden matures the pictures will be cheering to look at. When I go back to the beginning of the book now, I see that this kind benefactress Ann provided us with our first irises, the blue bearded ones, Japanese anemones, scillas, day lilies and a handsome thistle warningly called silybum. That was only the start, for during those early years she kept up a steady supply of things that did so much to help replace the rough grass and nettles in the yard. Sadly since then Ann has died but many of her plants, including a fine white potentilla, 'Abbotswood', remain as tokens of her helpfulness. Silybum for some reason declined. Ann Porges introduced us to a friend, another kindly, hospitable gardener who invited us to walk round her garden in both spring and summer, making lists of everything we liked, and in the autumn when the dividing of plants took place, we were invited to come to collect our share of them.

Early on we also met an engaging gardener who bubbled with eagerness for making lakes. On and on through his land he created them each year, filling the valley with water and the slopes with trees and naturalized flowers. He invited us over with orders that we should bring our own spades and sacks while he would provide a wheelbarrow; he then waved us off into his ten acres of woodland and grassy paths telling us to collect marsh marigolds and flag irises to our hearts' content. He also gave us some of his precious and dazzling white irises, *I. ensata*, which we now have growing in a group beside the pond.

These were just a few of the people who opened our eyes to a long chain of gardeners that zigzag their way round the British Isles, exchanging and giving plants to each other. Lists, seeds and planting advice must be forever on the move, forming a whole unseen sub-culture of botanical expertise. I like to think of these quiet lifelines reaching like small veins from garden to garden.

But by far the greatest influences on us – and they still are – were two sisters, Dorothy Hadoke and Maureen Thom. Both are singularly

knowledgeable, both are tireless in their own gardens, and each so generous there isn't a corner in our one and a half acres where we don't have one plant or another as reminders of their kindness. The first things Dorothy gave us, which have grown into successful and handsome plantings, were giant saxifrages, *Darmera peltata*, now in a mass by the water. They die down completely in winter so it's hard in spring to believe that their grotesque yellowish roots, the colour of the soles of old feet, gripping on to the bank are going to do good anywhere, but metamorphosis does take place. First tall stems with flesh-coloured flowers appear, these are followed by large glossy leaves with finely serrated edges which stand high against a view of the stream beyond. On the bank above is a rose, 'Paul's Himalayan Musk', which literally spills its massive branches in great arcs of pale pink flowers. This plant is a leviathan amongst roses; its prodigious growth spreads thirty feet or so in all directions and contrasts well with the saxifrage below.

Another early present from Dorothy which immediately upgraded the appearance of the yard with its handsome foliage was the royal fern, or Flowering Fern, *Osmunda regalis*. An impressive fern, being not so feathery as many others and with a frond of brown flowers. She also gave us an invaluable ground cover plant, an undramatic thing with the charming name of pink purslane, *Montia sibirica*. Hellebores, which Dorothy grows with the inconsequential ease with which the rest of us grow buttercups, were more early transplants to our garden. What mysterious flowers these are, coming into bloom in the nadir of gloomy days, providing great patches of subdued colour with their creamy petals, freckled with crimson and lime green – *Helleborus orientalis*, *H. foetidus* and *H. corsicus*. The latter have a wide branching growth with three-lobed leaves and flowers up to two inches across. Around our pond she gave us the orange musk, *Mimulus cupreus* 'R.C. Leslie', and the tall red 'wild musk', *M. cardinalis*. These I planted to replace the original musk, *M. guttatus*, which I had so mistakenly put in. They were far too invasive for the yard, behaving like ardent squatters with their tall yellow growth infiltrating up the sides of the pond, overbearing the washed colours of astrantias. Gradually I have forced them down the banks of the stream, where they can populate the margins at random. Even so, in spring I still find slyly groping roots trying to get at the primulas round the pond.

Primulas are another whole world of creations that should embellish every garden. What beauties they are – both *Primula sikkimensis* and the very tall 'giant cowslip', *P. florindae*, with their yellow flowers and the magnificent candelabra types, *P. japonica*, which take over the others with such hungry passion that Dorothy advised me to plant them downstream rather than round the pond. Their variety, length of flowering and diversity of colour make them paragons amongst plants.

Scattered about the place for their multiplicity of foliage and handsome presence are bergenias and hostas. *Bergenia cordifolia, B. purpurascens* or 'Ballawley' and *B. crassifolia*, are most agreeably natured when it comes to being moved about or divided. Amongst the hostas uprooted from Dorothy's garden are *H. sieboldiana*, the white-rimmed *H. sieboldii* and the waxy-leafed *H. undulata*. Rodgersias, willow gentians and cheddar pinks flutter their summer colours about here and there. In addition to all these she gave out dollops of encouragement and the best advice to any novice: to plant densely. If you do, not only is there no need to stake your flowers, but that 'stingy garden' attribute, when earth is visible in arid terrain between one plant and the next, is done away with.

Maureen also gave us plants with which to fill our spaces. Two evictions from her garden where she needed room for other things were *Rosa hugonis*, such graceful shrubs with tiny leaves that give a fuzzy outline showing off their yellow flowers, frail as porcelain cups. They proclaim the beginning of our rose season, reminders of the innumerable things Maureen has given us: astilbes with muted colours whose ferny leaves appear bronze in spring; *Iris sibirica*, diervilla and stephanandra. An upright shrub, *Buddleia globosa*, with scented flowers like round orange sweets, might just survive if the winters are mild. And even Maureen can't identify one of the plants she gave us, a showy broom, *Cytisus*, with its remarkable coppery flowers which look like Victorian chenille from the distance.

One plant above all astonished me. When it appeared for the first time I could hardly believe that such an exquisite thing could emerge when the earth was still cold – *Pulsatilla vulgaris*, the Pasque flower. I felt its pellucid amethyst flower and ferny soft leaves should be out somewhere in the wilderness, growing along hedgerows or sprinkling the meadows rather than in my miserly group where it isn't content enough to colonize. Finally Maureen's kaffir lilies, *Schizostylis coccinea*, are some of the last flowers around our pond; they go on blooming their heads off well into November. Walking with her round her own garden you almost have to restrain her physically and suppress your words of admiration, otherwise she'll gather up the seed of, dig up, break off or slice in half anything you remark upon before she can be stopped. Her generosity is boundless. Going home the car is loaded with boxes of little precious gifts or with fat, lusty roots of some four-foot beauty whose name, by the time I'm back, I've completely forgotten. I have to ring her up to remind me also of where it should be planted.

There is no doubt that without these two sisters whom we met through the osmosis that one gardener seems to have for another, we would still be in the dark ages, struggling to grow hibiscus and plumbago – legacies from other countries, other climes.

Michael and I do none of the things that gardeners do. We don't put cuttings into pots or seeds into trays; we do no propagating, potting-on or thinning. Our conservatory is for human beings primarily, plants are secondary. So when we had a busload of gardening society enthusiasts to visit us on a summer's evening, our garden was, for them, a dead loss. Their disappointment was apparent. They had hoped for unusual plants to identify, rarities to recognize, cries of 'What a good form!' to be tossed hither and thither, and naturally cuttings to be asked for. Instead they were faced with a late July shambles in which overblown roses were behaving with blowzy vulgarity, their raddled petals lying in extravagant pools.

I know that gardeners like these are more intense, more sincerely dedicated, more uncomplaining about labour and certainly more devoted to the vast commitment of the subject than those of us who flit and hope. They don't seem to joke – not about gardens, anyway. I call them the Penstemon Ladies because I once observed a group of Hardy Plant Society members walking round a garden where they were discussing those ravishing flowers. Pettifogging their way from clump to clump, their eyes seeing subtleties and refinements that had passed me by completely. Where I had been looking at a heavenly rampage of mottled and freckled flowers in a herbaceous border, they had been looking at a conundrum. 'Is that penstemon "King George" or "Schoenholzeria"?' 'Oh, it must be "Evelyn", surely.' Would one of them at any moment mutter fitfully about that winey-blue beauty known as 'Sour Grapes'? I moved forward hopefully. Discussion followed, animated and intense, and like a foreigner I couldn't grasp the language.

This attitude to gardens explains a lot. It accounts for the diversity there is; how there are some places which are wholly a collector's garden, where there is something of everything even if it has to be jammed in so that it's there, rather than precisely chosen because it was the only right plant for that position. Yet there is the other sort of gardener who anguishes over a decision; who carries a sample of colour, looking with fastidious care for exactly the right plant for a precise site, and the end result is manifest. Such a garden is harmonious and subtle, and appears to have happened with ingenuous ease.

Visiting gardens is a recurring pleasure; unfortunately for the majority of us it takes place just when we are most needed in our own. Visiting, though, is invaluable. We get so many ideas and so much stimulus that it becomes an important attribute to the business of gardening. On the converse side there is the pleasure of receiving visitors. Twice now we have opened our garden when other people in the village have opened theirs. Of the hundreds of people who have come through, each has seemed delighted to be there. They've stopped to talk, to ask questions and to tell us of their own gardens. They've sat about in the orchard, by the stream or in the summer house, and when the last visitor has finally left after seven in the evening, we've found no debris, no damage.

Walking through other people's gardens has become a national pastime. Many of the visitors here were obviously addicted to it, almost professionals. With heroic fortitude they were steadily working their way through places within a forty- or fifty-mile radius of their homes, loving every visit regardless of dimension and of weather. In the end, if year by year there is an increasing population spending hours in each other's gardens, what can it do for the famous places? Will it influence the head gardener's plans? Does a car heading north from the Elephant and Castle change next spring's plantings at Kiftsgate or Bodnant? Must paths be widened, steps be made safer, grass paths be replaced and shrubs be held back? Above all must there be teas? Gardens and teas are symbiotic, as inseparable as roses and black spot. There seems to be no end to what these jaunts have started; if it's accepted that no one goes in for purposeless strolling in their own domains then, of course, it has to be done elsewhere. And visually people can be an asset in certain places; they don't look like flowers, I know, but they do add a mobile dimension to designs that are supremely formal. In the Mogul enclosures, for example, Indian figures provided magnificent colour and animation to the precision. In the Boboli Gardens behind the Pitti Palace in Florence, the rather austere dignity is invigorated by the sight of pedestrians moving and children playing.

Walking round our garden with friends is another, different pleasure

because some of them are interested enough to be critical. Several of the best alterations we have undertaken are due to a friend who has made pertinent suggestions. His fresh contemplation has seen things we had grown used to, no longer having that wide vision we had at the beginning. We needed his discerning eye where ours had become accustomed to mistakes – his cogent remark that if we remove a short scruffy hedge the garden would spread visually to the horizon. The low box hedge which surrounds our terrace and provides formality in contrast to the rest, was instigated by someone else's perceptive eye, and again it needed an outsider's judgement to propose where we should add crucial height by building up an existing wall by several feet.

We are thankful for these imaginative comments because our approach to gardening is tentative and speculative. When someone stands around ruminating and then makes some sentient observation, it often solves instantly what Michael and I may have been pondering for months. As the seasons pass, recurring slightly differently each year, as gradually we acquire plants and knowledge from other gardeners, as we unpick our numerous errors, we have learnt one thing the hard way; it is this: take other gardeners' advice on design, or accept their plants, only *after* you have seen their gardens.

If you ask most people how they began their gardens they say they evolved. Few admit to sitting down and carefully working out the design. The majority seem to work gradually, letting things become apparent in their own time, and often at moments not knowing quite what they are dealing with.

I have read Russell Page's *The Education of a Gardener*, full of wisdom and stimulus. I love every word of it but I doubt if we have done one single thing he advises. I can sigh over the formal water channels flanked by conical yew trees in Jennie Makepeace's garden in Dorset, or David Hicks's immensely long rose pergola, and realize that before any of this is achieved somebody has to sit down with a pen and a piece of squared paper. It is no comfort to know it is possible, that really there are some people who do this. For the uninitiated it is hard enough to imagine that a three-inch cutting will eventually grow into a *Viburnum plicatum* 'Mariesii', a fine shrub with splayed leaves and creamy flat-faced blooms. Or even more of an enigma is a seed held in the palm of the hand. Is it really going to become a hollyhock, a six-foot stem in need of staking, with yellow and pink flowers evoking cottage gardens? As a child how agonizedly poignant I used to find the sentimental verses of E. Nesbit:

> Little brown brother, oh! little brown brother,
> What kind of flower will you be?
> I'll be a poppy – all white, like my mother;
> Do be a poppy like me.
> What! you're a sun-flower? How I shall miss you
> When you're grown golden and high!
> But I shall send all the bees up to kiss you;
> Little brown brother, good-bye.

Poppy or sunflower, at the start it didn't matter which, we were trying to cover the ground. Only later did I realize that starting a garden is the beginning of making a series of mistakes. Books explain how and when to do things, but they do not underline that one intrinsic pitfall that what you are about to encounter will never be concluded. There is no 'The End' to be written, neither can you, like an architect, engrave in stone the day the garden was finished; a painter can frame his picture, a composer notate his coda, but a garden is always on the move. Even Russell Page doesn't warn you of the elementary, in-built hazard that this thing once started will never be quiescent; at no moment, however peerless, will a garden stay immobile, petrified at its summit of flowering. And though I have now learnt that a garden is always in the throes of becoming something else, I still haven't come to terms with it. Forgetting, I make steps, plant plants, stand back and think the shapes work. Mounded cushions of saxifrage, all pretty pinks and whites in spring, soon become eiderdowns, making the steps impassable. Docile grey Snow-in-Summer, placed in becoming blobs of low growth, lets fly in no time and utterly overwhelms a choicely contrived arc of stones. In ignorance while I was planting, my mind had been on the present effect; as I tucked up a flower I had forgotten about the future. Even after several years of gardening, it is still pure chance whether I get it right with the placing of something or whether I must spend hours in summer hacking back a robust plant, too vigorous by far for the scant area I've allowed it.

Through our errors Michael and I began to learn the habit of plants; slowly we began to know the way each one would grow, the lie of its branches if it were large or the way it used ground if it were small. Every plant has its own way of filling space, but without understanding this

beforehand, how easily destroyed are those spaces which make the difference between a garden that sings and one in visual discord.

Our mistakes proliferate. A major one was thinking too small. Where we made curving steps down to the stream, they have been too diminutive, for we have to watch our feet when we should be watching wagtails; where we have made a pear tunnel it takes a few meagre paces to be out the other end, barely leaving time to relish the blossom and intriguing form of the trees. Raised beds are not raised enough to make a feature of their height, and where we have a path round the pond, it is not broad enough. Rather than being able to saunter and look, you have to totter perilously, thrusting aside leaves of Solomon's Seal and filipendula.

Then carelessly our other paths failed to begin at the beginning, indolently we let them evolve by following the ways we had naturally taken. We should have been resolute and designed the paths, fixing them like ley lines to hold down the overall composition. We realized too late why garden designers design. That mathematical precision between clutter and proportion. There is a right way where the size of garden, the dimensions of the house, the width of path and terrace, come together into perfect cohesion.

But how hard it is to get the arithmetic right between smallness and emptiness; and how perfect are straight lines. Why aren't they used more often? Is it because they are uncompromising? Because their rigidity somehow restrains the spirit, is inhibiting or unsubtle. Straight lines are good things; they are the direct route to perspective, to geometry, from the gate to the cottage door. Once these strong, decisive lines are fixed, defining boundaries, forcing direction and relating to the verticals of house and wall, only then can one achieve spectacular plantings and indulge in outlandish idiosyncracies. Endless patience is also required when planting huge-headed trees standing high, at right angles to the ground lines, where mounded shrubs or flowers soften and blur the crossing axes.

It is quite obvious that the horizontal view is not the only one: flower-beds, parterres, knot or sunken gardens, bedding plants and alpines, in fact all the looking-down on a garden is only one aspect. I was slow to realize how vertical divisions are paramount. How they lift the eye and intensify distance. They draw lines together so that what is planted below is framed as the focal point; or in the middle distance they can confuse an ugly outlook which needs obscuring. Upright shapes lay shadows over flat ground, making dense or dappled shade, giving an effect of movement or an illusion of substance.

However one thing we did get right, quite fortuitously: we can walk through the garden and return by a different route. It is essential, however small it is, to walk 'round' a garden, not merely to the end and back. So we have bridges and stepping-stones across the brook, steps to other levels, and mown paths cut at different heights which form a sort of architectural grass garden. Michael leaves some grass long until it needs to be scythed, when the wild flowers are over; some is of medium length, and some cut close to make paths that wind and cross each other, so that the levels of green in themselves make a design.

I know a garden should be reassuring, it should not unbalance your equability; yet because everything takes place in slow motion it may take years before arriving at the point of departure, the point at which things begin to go right. I have bursts of thinking we are there, that really each area is at last beginning to form its own identity; yet at the same time I can't stop the maddening habit, as I walk around, of mentally lifting things from one place to another, and imagining they would look so much better elsewhere. I suppose the great gardeners never went in for this twitching at the fabric; they had bold confidence and knew exactly how to combine structure with plants, and then keep it that way for the next half century.

Long before then, with a sense of urgency, I make a supplicating wail for country gardens to admit the country. Where is the charm in polystyrene? In those ubiquitous urns, so clinically white, which are placed on either side of the door; in those awnings, that mushroom lighting and umbrella-shaped clothes line; or in patios crowded with white garden furniture? Nettles, yarrow and ragged robin have been superseded by precast staddle-stones; the beautiful colours of rust and lichen are eradicated by doses of Solignum or Agrichem. Incisive grass edges are regarded as a prerequisite for any self-respecting gardener, while good behaviour is demanded from columbines and gilliflowers.

Walled gardens, behind their formal barricades and enclosures of architecture, are superb, but what about those small burgeoning plots of land held down by hedges of hawthorn, where crooked orchard trees and little bouts of dishevelment have all evolved through carelessness? Look over the hedge. Shouldn't a garden be related to all that? To the meadows, trees and lack of symmetry? Our villages are becoming too dapper by half. As Dawn MacLeod describes in her book, *Down To Earth Women*, an ideal country garden is a place 'which has slipped into the country scene without any dividing lines.'

Let's accept random seeding, let's tolerate small flowers like grace-notes decorating the paving. Not always, but sometimes. The corrosive vice of trimness infiltrates everywhere. Formality is pleasing in parks, imperative at Versailles and restful in courtyards, but why not let country churchyards be sanctuaries for wild flowers? A soothing sense of seclusion can still be found in churchyards, from gravestones, graves and grave-mounds, slumberous trees, ancient and umbrageous, and oak lych-gates, silvered with age. Here just the right balance between order and disarray can be found. Irish yews flanking the way to the church; grassy paths, mown and smooth, contrast with the untouched wild grasses full of snowdrops in winter, and later followed by celandines, dog violets and sorrel, meadow saxifrage, red clover and willowherb. How destructive, like wanton vandalism, to find some patches of churchyard wilderness utterly massacred by the rotary mower, or a wholesale pogrom carried out with weedkiller. As a result of these massive regimes of orderliness wild flowers, moths, ladybirds, crickets and lacewings vanish, followed by bats, wrens and hedgehogs.

Sometime soon we must reinstate our country churchyards and country gardens before it's too late. Telling the time by blowing the seedhead of a dandelion, holding a buttercup under the chin to see who likes butter, using a dock leaf to ease the sting of a nettle, will be lost to children. Daisy-chains, the drone of bees or watching a red admiral butterfly open and close its wings in sunlight, will belong to past childhoods. We will have destroyed our country gardens while looking the other way.

We are surrounded by low hills, so it is only necessary to walk a little way up our six-acre field to where we've planted our two-acre wood, to get a distant view of the village, the church, the cows and the sheep grazing in the small, medieval fields and the wild, bracken-covered hill where larks sing throughout the summer. Without the stream we'd be without frost, instead we'd be higher up and have a wider view. I've often wondered how we would deal with a garden which had a distant outlook.

The first time I discovered the intensifying effect of limiting a view was at a monastery in Greece. The monastery was on a hill and the courtyard totally enclosed by solid doors and a high wall in which there was one circular hole. Looking through it the impact was instant. From walking past the Byzantine apse and cloisters, unexpectedly there appeared a pristine view like an early Italian painted landscape seen over the shoulder of the Virgin Mary. The land fell away below the monastery to small pastures and a trail of dark cypress trees, meandering like leisurely mourners down the opposite hillside. The view was enhanced in a way which never would have happened if there had been no wall. The same effect was achieved in Hans Christian Andersen's fairy tale *The Snow Queen*, when Kaye, bored with winter, amused himself by making a peep-hole with a warmed coin placed against the window to look at the wintry roofs and falling snowflakes.

Since seeing that monastery courtyard in Greece I think that if we had a view it should be conserved by an enclosing wall or a tall, dense hedge. There would be circular windows or arches so that the landscape was observed deliberately. It would not be overlooked because it was an appendage to the garden, or because the plants at our feet were of such intensity as to leach out the delicacy of distance, but would be seen rather as a series of *trompe-l'oeil*, suspended like pictures so that the far horizon was neither forced nor taken for granted, but something intentionally looked at.

At Hestercombe in Somerset where Gertrude Jekyll and Sir Edwin Lutyens designed that magnificent garden with such bold use of flints, stone and cobbles, there is a wall with a circular window in it framed by roses. You look through and out of the fine garden to an undramatic view of pastoral countryside.

In America this result of inhibiting the view is to be seen at the Moon Gate, built around Korean tomb figures at Abby Aldrich Rockefeller Garden at Seal Harbour, Maine, in 1926. It is a very patent example of this *trompe-l'oeil* effect. A circular window, designed by the American landscape architect Beatrix Farrand, looks out from such a dark interior that the startlingly brilliant garden has an immediate theatrical impact. I hope somewhere in our English gardens there grows a deep canary yellow forsythia, 'Beatrix Farrand', with its exceptionally large nodding heads, in memory of this distinguished American lady.

I know nothing about American gardens, but I love to imagine what they are like. To us, on our little congested island, almost sinking under the weight of gardens, the idea of that space, remoteness and limitless horizon, where it is possible to walk over one uninhabited hill and find another just the same, is fairly mind-blowing. Thoughts of gardening then take on a different resonance.

I have been addicted to westerns for years, not for their prescribed formula of the good guys and bad guys, but because behind all their cavorting, often superbly photographed, is the most beautiful landscape

imaginable. Those ravines and cloud shadows across prairies; those black hills, menacing and remote, those distant mountains with snowy summits; swamps, creeks and gulches are words so irresistibly evocative I like to savour their potency, though I have little chance of ever seeing such places. High pastures, bone-dry river beds and contorted trees, misshapen forever by prevailing winds blowing constantly from God knows where. Think of it! An unimagined land-mass stretching boundless to the oceans.

'Land-mass' – what a heady word. Those of you who do not live on an island can have no idea of that sense of freedom when you set foot at Calais, aware that there is no watery restraint between you and Vladivostok. Little do the French appreciate that Calais is the gateway to emancipation. Fortunate Danes, Portuguese and Walloons. Enviable Americans, with frozen wastes and sweltering forests; with rivers like arteries passing through state after state. As for their gardens – how they vex me. I've tried to find out about them from our small local town library, but with no success. There is not one single book on American gardens. 'Well, we've never been asked,' explained the librarian. Isn't it time then, that we had a little cross-fertilization?

In our country, limited by climate and congestion, to me the idea of American gardens remains undefined. The extremes of garden design and what is possible to grow there, is fascinating to English eyes conditioned by roses. Given the diversity of nationalities in America, sources for inspiration must have come from five continents. Who gardens say, in Butte, Montana? What gems of wonder lie undiscovered in Meridian, Mississippi? Aren't there unacknowledged marvels in Bismarck, North Dakota; in Maine and California? And that state with the most beautiful name, Pennsylvania (a word whose first syllable should be uttered on an indrawn breath – while sylvania is released like a sigh), has, I'm positive, places worth leaving home for.

One of the most bizarre traditions that immediately strikes someone from England, brought up on privacy, is the American front garden. Where we have walls, fences or hedges defining our territorial frontiers, in small American towns they appear to have nothing. Where we clearly know whose leaves are ruining the lawn, or whose right it is to pick an overhanging sprig of lilac, over there there are no definitions. Where did the open-plan tradition originate? And doesn't this system presuppose that every house owner is pleasantly co-operative towards his neighbour? Who mows what? Is there no animosity over shadows? What happens, for instance, when one person decides to plant a coniferous tree; when loose petals of overblown blossom float across invisible boundaries? And are you obliged to greet your neighbour as a *force majeure* that you meet eyeball-to-eyeball across that undefined sward?

Yet the effect of seeing a green flow of grass which unites each house one to another gives a restful appearance to the street. Trees and bushes are there, but not the individual expression from flower-beds.

At the back of the house, obviously it is different. Here is a description of a garden in Iowa, taken from Jonathan Raban's account of his journey down the Mississippi in *Old Glory*:

> . . . I followed him behind the bar and into a scruffy little kitchen where he switched on a raft of lights which flooded the yard beyond.
> 'Look,' said Mr Frick, showing me out through the door.
> Even now, his garden was still an embroidered quilt of summer colours. Plants in pots were arranged in steep pyramids, in banks of deep green ferns, in white, wrought-iron pagodas and hanging baskets. He had squashed what looked to me like a complete Chelsea flowershow into the space of a living room: slender herb margarets, livid begonias, fuchsias, chrysanthemums, primulas, geraniums . . . Tiny gravelled walks trailed in and out of the beds of flowers. The centrepiece was a miniature waterfall. Mr Frick switched it on, and a little river came bubbling through the ferns, over rocks of coloured crystal and splashed into a lily-padded pool. In the corner of the yard was a rose bower, the blooms of pink and crimson looking bloody in the floodlight. A signboard with carved rustic lettering was suspended over the top of the bower on silver chains: it said 'I Never Promised You A Rose Garden'.

One of the attributes of American gardens that is alien to us, is the way in which they can be taken into the house, or, conversely, the sitting and eating rooms can be part of the garden. From looking at pictures I see that many of these designs use pebbles, boulders, gravel, cement and driftwood, while plants in pots are used for twining up dividing screens. In Penelope Hobhouse's book, *Garden Style*, she describes a small garden in New York where pots of grasses and bamboos, as well as flowering plants, are used ' . . . like flower arranging; pictures are constantly built up and then changed; the overriding aim is to create a green oasis as a contrast to the concrete city desert.'

In the same book is a lovely photograph of the Magnolia Plantation in South Carolina. The still dark waters reflect the straight lines of the 'live oaks', with their roots rising like sinews to hold the trunks soaring high above the swamp. Are there still to be found legacies in the gardens of the south from all that available labour in other centuries? Are there fine gardens laid out around fine houses in the midst of fields of tobacco? Is Rosedown in Louisiana a ravishing place of monumental trees, azaleas, camellias and sweet olives? Did traders on the Mississippi make spectacular productions between the sugar-cane and cotton plantations? Perhaps the Missouri Botanical Garden in St Louis is a fantasy place, with desert houses full of night-flowering blooms and grotesque exotics. Cypress swamps, magnolia gardens and banyan trees; bayou and savannah – what foreign words to evoke a wild extravagance of sensuality, way beyond the quiescence of sweet williams.

As for the 'butterfly pools' in Charleston, imagination takes off. Are they migrating flecks of blue, drawn by an irresistible plant grown only in Charleston? Do the butterflies waver and surge and move like the sea in their breeding season? Elsewhere what colours are to be seen in the Longwood Gardens of Pennsylvania, among the rolling hills?

There's so much more, but I shall never know. There are, fortunately, wonderful American writers that do penetrate my sitting-room. That witty and abrasive writer Henry Mitchell does slap down my pretentions to try for the impossible. Succinctly he tells us that one bulb may be enough, or that there really are alternatives to 'a million blades of grass shorn uniformly'. And I have taken to heart his advice to 'not go hog wild' over peonies, much as I adore their plushy faces and sturdy stems.

Another much beloved book is Eleanor Perényi's *Green Thoughts*. She has a wonderful chapter on blues. Not on moods, but on flowers. 'I love blue more than any other color. I am inordinately attracted to any blue substance: to minerals like turquoise and lapis lazuli, to sapphires and aquamarines; to cobalt skies and blue-black seas; Moslem tiles – and to a blue flower whether or not it has any other merit.' Then off she goes beguiling us with her words and descriptions, her enthusiasm for and commitment to all the blue things to be grown in a garden. When she comes to platycodons I have to turn to my garden encyclopaedia, for shamefacedly I've never heard of them. They are balloon flowers and look from the photograph the most exquisite deep blue with papery-veined petals. The balloon-shaped buds open into saucer-shaped flowers.

In the book Mrs Perényi comments: 'That Americans aren't a race of gardeners is evident enough from the range of tools offered in most stores.' Maybe. Even so I'm not put off. I long to see the variety of their gardens through eyes that for years have been focused on those in England.

The third major influence on our garden, after trees and water, was walls. The houses in the village are built of stone, not of the warm golden colour of the Cotswolds, or the dark sombre tones of Wales, but of good grey with more yellow than blue in it. Our cowshed, granary, duckhouse and outlying walls are of this stone, which perhaps is amongst the best material for using as a background to plants.

We began by making a terrace about seventy feet long and twenty feet wide, reflecting the dimensions of part of the south side of the cottage. I love the simplicity of seeing the two planes of stone, the house and the terrace, joining at their right angles – we didn't want flower-beds under the windows, but merely left gaps for climbers. Into one of these holes, surrounded by almost achromatic stone surfaces went a rose, that lovely old-fashioned, free-flowering 'Gloire de Dijon', whose blooms were to be highlighted by a creamy clematis, the 'Duchess of Edinburgh' – both would contrast divinely with the intense azure flowers of *Ceanothus* × *veitchianus*. Our colour plan was good we thought; the tea-scented, many-petalled pale orange of the rose was to mingle enjoyably with the ceanothus. That combination of apricot and blue is one I find irresistible. It is in the traditional dresses of the French peasant figures painted on Quimper pottery; in a fragment of Egyptian wall painting from a tomb at Thebes; and inadvertently I've made it work by planting nepeta with potentilla 'Day Dawn' under a crab-apple tree. Every year they flower simultaneously and leave me with the same incredulity and satisfaction of having, quite by chance, got the colours right.

Alas none of this happened against the cottage. For not only did the 'Duchess of Edinburgh' come out long after the rose, in fact she's always in bud well into November, far too late for common sense, but the ceanothus was each year frosted crispy brown and the most we got from it were a few blue scraps in April, naturally long before 'Gloire de Dijon' had even responded to spring. The colour scheme foundered before it had started and anyway, after three winters, the ceanothus died completely. I'm still trying though. I reckon to make three attempts before surrendering, so now we've replaced the dud ceanothus with *C. impressus*. It's surviving by being

nannied in winter, when it's covered with a blanket of bracken, held in place by netting. But really, is this the way to garden, I wonder? Shouldn't I submit, shouldn't I wisely accept that we live in a cold climate and only grow those things that adore a bitter east wind, icy nights and frost in early June?

Further along on this same wall we planted the rose, 'Zéphirine Drouhin'. It is not the best of places, but she loves it. Every summer the stone is totally obscured by a massive outburst of magenta flowers. There is no subtlety about this rose. Any poor thing within the rays of pink which engulf the atmosphere with scent and vibrancy has no chance. For five weeks or so the rose is supreme in this part of the garden; all the silvers and greys, the mild blues and faded yellows can be seen only if we deliberately turn our backs on this most outrageously prolific rose. We love her dearly, and bless her sumptuous generosity in giving such a glut of colour from a twelve-inch hole, under which we know is the revolting puce clay on which the terrace is bedded.

At the beginning, when surveying our many walls, roses were some of the first things we went in for, and on the whole they turned out to be tough, compliant and indifferent to aspect. But there are many other plants, tempting and challenging, which should benefit from growing against a wall which retains warmth and where, after all, the wind damage isn't so calamitous, and snow can't overburden the boughs as it does with a free-standing shrub.

Having assumed our north wall would be fatal to plants, we could hardly believe what books told us; the list of things recommended for a sunless aspect was unimaginable. Following such sanguine advice we planted a morello cherry; fan-shaped and quick-growing, it has spectacular snowy blossom every year, followed by a brilliant crop of cherries, half of which we net to ripen and which end up in jars of brandy, and the other half we leave for the birds who seem besotted with them.

Beside the cherry is a strange, rather reticent rose, 'Paul's Lemon Pillar', with acid-yellow flowers which smell of aromatic tea. Then famous for thriving on a north wall is that magnificent and obliging climber with creamy flowers, *Hydrangea petiolaris*. Not only is it rampant on this cold aspect, but it is also self-clinging. What a beautiful description, 'self-clinging', to find written in a catalogue. It does away with hours of fiddling work, fixing up wires and supports, and this particular plant really is supplied with the most effective sessile system. Millions of little fringy legs hang on with as much tenacity as a limpet to a rock. How different from

Parthenocissus quinquefolia, the 'Virginia Creeper', which we have on the cowshed and which is temperamental about its clinging habits. Sometimes it does, sometimes it doesn't, and I can never fathom its reasoning.

Next to the hydrangea is our pesky rose, 'Conrad F. Meyer', a plushy Rugosa, but hopelessly truculent when it comes to being fastened over the back door; whereas beyond it, festooning an iron arch we put up to support its luxuriance, is an early *Clematis alpina* 'Pamela Jackman', with nodding flowers, which forms a long blue tangle surprising us that anything can look so summery while it is still April. But I do wish clematis didn't have such hermaphrodital tendencies: they will cling to themselves when I want them to spread their affection elsewhere – to nearby climbers, preferably. 'Madame Edouard André' has this self-loving bent. She is meant to be pouring her dusty red flowers with flagrant abandon into the embrace of a nearby escallonia, but nothing comes of it.

Walking down from the back orchard towards this part of the cottage provides us with visual proof of the authenticity of advice. These plants along the north wall – and I haven't named them all – really do grow with a continuing array of varied flowers and leaves for months on end, even though, for other months on end, they endure the malice of vile winds and the rigours of glaciation. We know now with growing confidence that the softening effect on any building from the contrast of colour and texture is something to be endlessly worked over from the very beginning.

What a pity more house walls aren't covered. Resilient clingers like *Hydrangea petiolaris* won't do, I realize, for houses that have to be repainted, but there are other more amenable plants that can be pulled aside. For how well plants diffuse the red of bricks, the disagreeable texture of stucco or the bleakness of concrete. Are people afraid? Afraid that snaky tendrils will creep under tiles or insinuate their way into drains? Do they fear the mess, fallen leaves or bird droppings? In France the hackneyed virginia creeper is everywhere. How well it suits those large square bourgeois houses or the long walls of an inn. Twenty years ago a dreary little housing estate was built very close to Corfu airport. The houses were small cement squares, built with relentless tedium and economy to the eye, and yet within a summer's season the effect had been transformed into a garden. It was all one muddled prettiness from vines over hoops leading to the doors, jasmine and roses over the porches, to pots of geraniums on the doorsteps. The front gardens overflowed with octopus arms of melon plants and hummocks of leafy courgettes with their yellow trumpet flowers; jumbles of flamboyant peppers and tomatoes supported on twigs grew in a cosmopolitan chaos amongst some lilies. The result was magic: the bleak boxes had become stuffed to overflowing with lavish prodigality, practical and unaffected.

On our walls here we found that honeysuckles were a boon. They seem to mingle so benignly with anything, whether it is the original hedgerow Woodbine, *Lonicera periclymenum*, or a rarer, more startling honeysuckle with modest vigour and resistance, a *L.* × *brownii* 'Fuchsoides'. This lonicera is in a sheltered corner, its scarlet flowers borne in whorls, though not yet with the grand munificence I keep hoping for.

On the shorter length of the west wall are more roses, clematis and honeysuckles: the pure white 'Iceberg', the late *Clematis viticella* 'Etoile Violette' and a showy *Lonicera* × *tellmanniana*. It may sound unadventurous that we repeat the same sort of thing round and round the house, but it wasn't always like that. At the beginning we gathered in every desirable and possible climber that lured us by its description: *Stauntonia hexaphylla*, an evergreen twiner with small fragrant flowers of white tinged with violet, *Ercilla volubilis*, again with purplish-white flowers in spring, and *Eccremocarpus scaber* with bright scarlet tubular flowers. Each in due course passed on; they hadn't a hope of ever reaching maturity. *Schisandra chinensis* might still make it if we have a few kindly winters, but the sweet-smelling starry *Jasminum officinale* needs comforting, so I'll try it in the conservatory. That is where I have the impossibly named *Trachelospermum asiaticum* which, with its dark leaves, is supposed to produce hundreds of ivory flowers with yellow centres. Poor thing, pining for Korea, it hasn't produced the trace of a bud.

These are just a few plants that have snuffed it. My garden book is full of burial crosses beside each entry. Yet elsewhere against the walls other things grow as easily as pie. Along the lane we have built up the wall to protect us from the east wind. Here is a chocolatey-coloured twiner smelling of vanilla. In my cookery book entirely on chocolate the author, Helge Rubinstein, writes one chapter of such rich concoctions; it's called 'Death By Chocolate'. One sniff of this plant, *Akebia quinata*, and the garden vanishes in a haze of gluttonous memories.

'Mrs Cholmondeley', a blowzy clematis who draws scoops of light, even on dull days, to her intensely blue petals, grows against a wall where the trespassing coils of *Lonicera* × *americana* reach to her from around the corner. This heavily scented honeysuckle, full of bees amongst its whorls of deep yellow tinged with purple, is rooted next to a rose of unbelievable beauty. 'Mme Caroline Testout', whose strong branches we have tried to train in arcs against the wall, has globular flowers of satiny pink. With her is an early *Clematis montana* 'Tetrarose', whose serpentine spirals are our annual despair as we try to cope with its intricate network.

How unpredictable plants are. The soil where once there was an entrance to the stable is impoverished, stony and thin, yet that east wall presentation is nearly always worth a second look, when elsewhere, in a carefully composted mix looking as rich as Dundee cake, the rose 'Albéric Barbier' never lets fly with yellow buds opening to creamy clusters.

That's another thing – the vagaries of flowers. For why can I never grow aconites or meconopsis other than in stingy handfuls? Why won't nerines or pulsatillas abandon themselves in coloured clumps about the place? Other people have these pink or purple swatches. Where have I gone wrong? On the other hand there is 'Mrs Cholmondeley', wantonly letting herself go year after year with huge, regenerating blooms of unadulterated blue, in spite of the matted wreck she has become, and with which I'm quite unable to cope. How can she like that disarray? The one year I did attempt to tidy her up was disastrous. She remained prim. I shan't try that again.

I remain baffled; there seems to be no fine line dividing right from wrong as to how to treat a plant. It either takes to you, or it doesn't. Following the book, going through the procedure, I can see makes sense, but there is another element at work, and that is the waywardness of plants. Long may it remain like that. Never let the garden centres come up with foolproof plants.

When we first started I had an instinctive longing for a garden of stone. Of course I don't really mean just that, a bald stringency of stone, but long before plants were to go in I wanted to contrive structures of stone. It wasn't easy because we only had a limited supply, gathering it from here and there as we went along. We are still devising places of stone, even though ideally this should have been done from the start. Stones and foliage, the shock or balm of colour, surprise and repose around corners, were each intended to weld harmoniously to make a delightful garden.

It doesn't really matter what the solids in your garden are – whether wood, brick or stone, depending on the part of the world you live in – but it does matter that they should be there. Wherever you see a garden with strong lines of construction – sometimes from ironwork gates, pergolas or railings – they immediately add the perfect complement to plants. Brick paths in patterns, cobble stones or paving intensify the elegance of flowers. Whatever they come from, the contrasts must be tangible.

Living in this part of England, where limestone breaks through the earth as easily as the bracken does on the hill, we are surrounded by folding and subsidence pre-dating the first dinosaurs and flying reptiles. Somewhere around seventy million years ago, I suppose, this stone I've laid down was a cephalopod or graptolite living a placid Palaeozoic existence, quite unaware that its immortality would be remembered as I make a wall for the herb garden.

My addiction to stone originated in the Mediterranean countries, where in certain areas the land under olive trees has great outcrops of greyish-green rocks; a whole grove echoes this glaucous neutrality and reflected light from the silvery foliage. Stand below a grove and look up at terrace beyond terrace of dry-stone walls throwing back the cohesion of turbid colours; in the ceaseless rain of winter the trunks of the olives turn black, and only a tethered sheep or donkey animates the scene.

It was with this image in mind that I terraced the back of the house where the land rises steeply. Rather than let it be a slope planted with flowers, I started one winter with a low stone wall about seventy feet long and about twenty inches high. At different levels further up the bank I made small curved walls of differing lengths, providing flat planting areas. The informality of these secondary walls allowed for asymmetrical design, where low-hanging plants could drape from level to level, but not so prolifically as to hide the stone construction, which was intended to be as important as the plants themselves.

But I'm disgruntled: I should have done it on a large scale. With giant strides I should have marched around and then I might have realized that to get the drama from a design like this needed boldness, not the niggardly fiddling which I went in for. As a kind of intriguing little garden in the shadow of the house, it could be something, I suppose, but it was not what I had intended. Damn the vigour of plants. Once again I forgot to allow for their growth. I didn't really want to be landed with places for microscopic plants, little precious treasures to be bent over and peered at; I wanted a dashing effect of terraces with a pushy appearance, scenic and assertive.

Stone is everywhere in our garden. We've laid huge spans of it where steps lead down to the brook. Here the water runs deep, so one good, wide stride takes you over to small steps leading up the other side. Elsewhere we have a paved terrace backed by a stone wall, fat and low enough to sit on. In the centre of this terrace I've made a 'pond' - a circle of broken crockery collected from all over our land. Household rubbish must have been dumped outside for generations, slowly sinking into the earth, but never being entirely integrated so that even now, whenever we dig, we are bound to come across a bit of cup or platter. These archeaological finds are the history of the place; rather than throw them in the dustbin I collect them. My pond is made up of pieces of slipware, spongeware or willow pattern; fragments of glazed pinks and purples, but predominantly blue, set into a cement circle. It makes a pretty pool of colour beyond which grows lime-green euphorbia and an amber yellow rose, 'Maigold'.

There isn't a time when we aren't contriving something out of stone. Plans get carried along by a headful of resplendent ideas based on the complicity between stone and plants: shadow, substance and colour. One year we made a sunken garden. Shallow steps lead down to a paved area where we laid large, irregular slabs once used as the threshold to stys and byres. On two sides of this stone garden, against the relentless westerlies, we have planted yew hedges. One day these will be tall, structured and dominant, holding in the aromatic heat not only from various thymes, camomile, penny royal and marjorams between the stones, but from cystus, gypsophila, rockroses and lavender. Balls of santolinas grow along the base of the yews in silvery precision, showing up well against the inky-green background. The tops of two ubiquitous staddle stones are upside-down on the stone floor, making sculptured blocks as well as providing somewhere to perch briefly, while snuffing up the redolent scents of what I hope, one day, will remind us of the Mediterranean.

Stones will go on being dug in here along with manure and flowers year by year. Three years ago I designed a stone loggia I want in the upper orchard; so far the drawing remains in my garden book along with plans for a ruin. I wonder how that would look with climbers festooning its decaying heights.

Gardening is not uphill all the way, but there aren't many days when I can confidently coast downhill with my hands in the air. Just when I think I'll give up, when, unlike my roses, I'm not feeling in the pink, and the whole gardening hierarchy is a bore, then I discover a group of campanulas the blue of thrushes' eggs inadvertently growing through our *Lonicera etrusca*. Quite by chance the mixture of these two plants, the unusually pale blue of campanulas and the buttermilk flowers deepening to topaz of the

honeysuckle, gives me fresh heart to keep going.

Amongst some of the continual garden boons, scents, sights, colours and good-natured plants that keep a gardener shored-up with optimism are clematis. We have them everywhere, about twenty or thirty varieties – we can never be sure because sometimes they'll flower willingly and then they'll disappear. Just when I think one of them has succumbed either to wilt or to frigid hoar-frost, I'll come across what looks like a stringy bootlace. Following it to its source I discover miraculous tender green shoots. Where have they been all this time? Seemingly dead for a year, lying low until one day they confound me. Another clematis bonus, unexpected and soothing, comes from untwining a recalcitrant tendril of the rosy 'Comtesse de Bouchaud', assigned to grow amongst wisteria. It is an infinitely delicate, infinitely painstaking task that brings calm to the soul.

Another benison comes from habitual closeness to certain parts of the garden. There are small corners, entangled with weeds, to which throughout the summer I have to return again and again. On my knees, delving between plants so closely that my breath moves them, I become familiar with the earthy smell, with the sheltered comfort of groping at the root of things. In this way compensations are showered on us. At a moment when I'm prepared to walk indoors, shut myself up and get excited over the Repeal of the Corn Laws or the making of dolls' eyes, I catch sight of the water-lily leaves. In summer wagtails strut about on these green discs, but on some day in autumn the leaves turn deep ochre, floating on the surface of a blue reflected sky. Or on a cold morning when leaves are falling at countable intervals and any day soon a frost will bring them all down, comes an instant uplift from seeing the dramatic explosion of red bergamots, like a last hot blast of summer. How often walking round the garden after frost reveals unexpected pleasures: the corrosive colour of a Welsh poppy still in bud amongst the leathery leaves, green and veined with purple, of a sage; one bright veronica still flowering; two remaining unseasonal dusty pink blooms of *Clematis chrysocoma*.

Gardens are always doing this. After weeks of myopic peering at our frosted bay tree, we discover a fleck of life. It's not just the Psalms that say it flourishes, it really does. And as the result of being given the wrong rose from a nursery, quite fortuitously we brought into our garden 'Celsiana', an eighteenth-century Damask with clustering pink flowers and downy green foliage. I love her to bits and now wouldn't want to be without her.

One final grand bonus that comes at the end of the year, never to be underestimated or taken for granted, is walking. Until you have spent several months crouched with bones compacted, vertebrae consolidated, arms strained to the ground from carrying buckets to and fro; until you have shuffled all the summer, inclined over a weighted wheelbarrow, tottered with tools across your sagging shoulders, until this has happened for far too long, you can have no idea what it is to go for a walk. To walk upright. To feel your back extended, those extra inches returning to your spine; to be aware once more of the utility of bones, the brilliance of mechanics, of a hip socket, smooth as a ball-bearing. There is nothing like the euphoria of rediscovering your limbs as individual swinging appendages after a long, bent summer.

4 The Will of the Rose

Botanical Illustrations / The Introduction of Etching / Parkinson, Redouté and Margaret Mee / Genocide to Buttercups / The Intransigence of Roses / A Folly with Three Roses / Feral Roses / Ruminating, Lolling and Lazing / Soulless Flowers / In Favour of Green Gardens

A HUNDRED AND FIFTY years ago Michael's great, great-aunt Elizabeth Ranyard painted wild flowers in a book inscribed in spidery handwriting. Inside, on pages of beautiful dense cream and sometimes coloured paper all edged with gold, are paintings of such minute exactitude that a stamen or thorn appear to have been painted with an eyelash. The pale mauve crumpled petals of an opium poppy are done with a light touch as frail as gossamer, each hair on the stem finely lined without emphasis. The purpley backs to the stems of coltsfoot are graded delicately, the colour merges into the green so naturally the flower almost stands clear of the page. The coarseness of the leaves of common comfrey which she found on the 'shores of the Thames' seems tangible. *Veronica chamaedrys*, smooth-podded 'Sea Vetch' and 'Lady's Finger Kidney Vetch', bird's foot trefoil, sea samphire, bee orchis, red hemp nettle and flimsy wood betony are among the flowers included in her book. The only concessions to artistic interpretation in some of the meticulous paintings, are very slight shadows that give a certain depth to the background.

Michael still has Elizabeth Ranyard's painting box. Looking at its bulky squareness it is hard to understand how these ladies of the last century went about their painting, unless, like Queen Victoria's granddaughter Princess Patricia, they had an escort of Japanese servants to carry their sketching implements. I somehow imagine aunt Elizabeth had no such help. Her walnut chest is full of graded paint brushes, some black lead powder in a small wooden box, various porcelain screw-top jars and tubes of paints from a wholesalers in Glasshouse Street. A wooden tray, divided into sections, each made to hold a rectangle of paint, lifts out. 'Indigo', 'Indian Red', 'Lamp Black' and so on are stamped on one side of the paint blocks, with the name of the suppliers, 'Newman, Soho Square'. Underneath the tray, wrapped in a piece of yellowing newspaper, is a glass plate for use in a magic lantern, and a little book of price lists dated 1840. 'These Outlines are prepared ready for Coloring, by means of which young people may paint their Slides. Price 6d each.' For another 6d there are 'Lectures to accompany Magic Lantern Slides', cloth-bound and gilt-lettered. Amongst the titles are 'Blue Beard, and the Timid Young Man', 'Pussy's Road to Ruin, and the Lay of St. Dunstan' and 'The Witches Frolic, and Lord Bateman'.

What talent in the drawing-room. How necessary the catalogues must have been, listing gilding and wax-flower-making materials, modelling tools, drawing pencils and *best* drawing pencils. What infinite patience and solitude are revealed in Elizabeth Ranyard's flower paintings. I imagine her living a safe, restricted life in which embroidery, music and water-colours filled hours of every day; in which ladies of the last century occupied themselves by working with painstaking self-discipline at something which, in the end, was shut away in a book.

Yet these paintings of flowers were not executed for the sake of artistic merit, as something decorative or sentimental. They were as botanically inspired as those done for medical identification. The observation from live plants, not from copying older illustrations, was exact and scrupulous. It was as a result of looking at Elizabeth's flowers that I began to wonder about the source of botanical illustrations. Who were the first to draw with such exactitude? What had been their motive, and how were the drawings done? The Victorian ladies were doing them for themselves, to take up the slack in their prescribed lives. But the origins of this form of art had occurred centuries earlier, for medicinal purposes. As long ago as the first century BC, Pliny the Elder gives evidence of illustrated herbals.

In the National Library of Vienna there is a manuscript, the *Codex Vindobonensis*, produced in 512 by the physician Dioscorides. From this herbal of almost four hundred full-page drawings, most of which are coloured, proliferated copies, adaptations and translations. But the most significant illustrated herbals began with a Carthusian monk, Otto Brunfels, born in the late fifteenth century. The draughtsman for his *Herbarum Vivae Eicones* was Hans Weiditz, an artist who has been compared to Albrecht Dürer for his fine and masterly work. Looking at a woodcut of a pasque flower it's incredible to see such vitality and detail, in spite of the non-fineness of the medium. When you consider that the wood is cut away from the drawing done on it, it's remarkable to find such precision, even in the hairs on the stems and the fernyness of the leaves.

By the mid-sixteenth century Fuchs, a professor of medicine in Munich, had produced his magnificent folio volumes comprising double the number of illustrations contained in Brunfels's work. And, like the plant collectors who have left us reminders of their existence in the form of the plants in our gardens, the memory of this field botanist is perpetuated by the many species of *Fuchsia*, though oddly Fuchs never actually saw this American genus. Christophe Plantin, a printer and publisher, also indirectly had a long-lasting influence on our gardens. In the second half of the sixteenth century he amassed woodblocks of plants, and produced books of distinguished Flemish herbalists. One of them was the great botanist, Dodoens, who travelled in France, Germany and Italy; another was the even more eminent Carolus Clusius (Charles de l'Écluse).

Clusius was a doctor and horticulturalist who not only observed and recorded plants in minute detail, but whose ardent efforts to cultivate plants from Turkey and the Levant have left northern Europe with a fine legacy of garden flowers. These include *Iris sibirica*, with its unpredictable variations of blues, and the charming family of *Primula auricula*. I find these sturdy-stemmed little flowers with their chunky shape difficult to place in our unkempt garden; they need somewhere confined to show off their Victorian faces, reminiscent of wallpaper and dimity. Later in his life Clusius became curator of the Hortus, the botanical gardens in Leiden concerned with ornamental and medicinal plants and herbs. To this post he brought his own passion and gusto for growing bulbs: irises, anemones, lilies, hyacinths, narcissi and, above all, tulips.

Tulips and the Netherlands are today synonymous to us all. They weld in our minds into a vast colourful production. It is to this indefatigable botanist, Clusius, that we owe a debt of thanks for some of the most beloved spring flowers in our gardens. The seeds and bulbs were first sent to him by his friend Ogier Ghiselin de Busbecq, an envoy of Emperor Ferdinand I at the court of Suleiman the Magnificent, Sultan of the Ottoman Empire, of which, charmingly, the tulip was the official emblem.

In 1597, John Gerard's renowned herbal, *The Herball or Generall Historie of Plantes*, appeared. This had very few original plates, most having been taken from blocks of earlier herbals. But the work spored innumerable manuals, including Mrs M. Grieve's, published in 1931, and now available in an immensely fat paperback edition.

A dynamic turning point came at the end of the sixteenth century, with the introduction of metal engraving. Precision and fine delineation were now possible from engraved or etched copper plates. In quite the opposite way from woodblocks, the artist's drawing can be engraved on the plate with a tool called a burin; the plate is then inked, the surplus wiped off, and the paper pressed hard on to the surface.

Etching, on the other hand, is done with a needle, after the plate has been covered with a wax coating to protect it when immersed in acid. The etching needle pierces the wax, allowing the acid to eat into the metal from which a print is taken. There is more freedom in a way in etching; it doesn't require quite the force necessary to push the burin. I have to admit, though, that when looking at a drawing of a plant, I cannot be absolutely certain whether it is an etching or an engraving. What is apparent in both techniques is that they provide a means of shading; depth of tone as well as texture can be depicted in these illustrations, adding another dimension to the clarity of the plant.

There followed a series of superb collections, such as Basil Besler's vast *Hortus Eystettensis*, which appeared in 1613, with almost four hundred plates, engraved by a number of artists. Wilfred Blunt refers to this collection in his fascinating book, *The Art of Botanical Illustration*, in which he describes an engraving of 'hyacinths and ornithogalum' ('Star of Bethlehem'). 'The designs are really impressive, and the invention rarely flags; the rhythmic pattern of the roots, the calligraphic possibilities of lettering, are fully explored and utilised; and the dramatic effect of the whole is enhanced by the noble proportions of the plates, . . .'

This was followed a year later by Crispin de Passe's *Hortus Floridus*, which contains about two hundred illustrations, some humorously embellished with an insect hiding amongst the leaves, or a mouse gnawing at a corm.

Skimming along through the countries and the centuries of consummate artists, I must mention two more before reaching Banks. One was Nicolas Robert, who was employed by the Académie Royale des Sciences to assume the responsibility for the illustrations of *A History of Plants*. It marked a watershed in the high standard and magnificence of this type of work. Finally, in 1788, long after Robert's death, the completed history appeared. And the other notable illustrator, with his optimistic middle name, was Georg Dionysius Ehret, who was born in Heidelberg in the early eighteenth century. He was to have an outstanding influence on botanical art; his output was prodigious, his paintings pre-eminent. Some can be seen in the Print Room at the Victoria and Albert Museum, and some at Kew. Having married the sister-in-law of the curator of the Chelsea Physic Garden, Ehret spent the rest of his life in London, being patronized by the nobility. The brilliance of his engravings led to his election as a Fellow of the Royal Society, the only foreigner on the English list. An honour which, much to his dismay, meant stumping up twenty-five guineas.

Now we come to an outstanding source of exquisite botanical drawings due to that great patron of plant collectors, Sir Joseph Banks. On his journey round the world between 1768 and 1771 with Cook, Sir Joseph employed Sydney Parkinson as his chief artist. From the huge collection of Parkinson's drawings, some of which were only sketches, and had to be finished by others after his death, are those of animals, landscapes, boat-buildings and his superb plants. Just to look at his illustration of *Pereskia grandiflora*, a non-succulent cactus with pink and white flowers and deep green leaves, is to get a vivid example of the man's work. Returning home after the voyage round the world, he died of dysentery, at the age of twenty-six.

After the technique of colour printing had arrived, there followed some of the most celebrated flower illustrations, including those of Pierre-Joseph Redouté. Unlike the subjects he painted, he seems to have been a repellent-looking man, with grotesque limbs and a distorted head. His *Les Liliacées* and *Les Roses* appeared at the beginning of the nineteenth century. Using stipple-engraving, a method similar to etching, except that instead of drawing lines with a needle, the latter is used to make dots, Redouté drew his portraits of roses. The stippling allowed for a greater subtlety, a more fragile difference in tones, achieved by minute variations of pressure on the needle. Most of the originals of the Empress Joséphine's roses at Malmaison are in Paris, but even in reproduction, the delicacy of this work makes the ancient rose 'Rosa Mundi', with her crimson petals splashed with pink and white, appear fine as glass.

Start chasing botanical illustrations and you begin to explore fascinating territory which meanders all over the place. I discovered, for instance, that officers of the East India Company taught Indian artists to make drawings of plants for them. These beautiful works of art ended up in museums and libraries in England, the plants annotated, the artists nameless. Using sepia ink, Indians were taught by their masters how to look minutely at a specimen, before drawing exactly what they saw. Abandoning their own more showy and traditional style of painting, they needed to learn a method of naturalistic precision. What legacies they left us. Look at a painting of a Mysore thorn, *Caesalpina decepecala*, with its luminous light yellow petals, separate diagrams of stigma, sepals and stamens, seed pods open and closed, and you see just how effective their teachers were, and how equally versatile the artists.

Facing a different direction, and following another thread of flower illustrations, leads me to *The Botanical Magazine*. This was an impressive and innovative publication founded, in 1787, by an ex-apothecary, William Curtis. In spite of its erratic career, the magazine is still going strong after two centuries, and has been described as 'the oldest horticultural periodical in print'. The first issue, in blue paper covers, included three hand-coloured engravings, with notes on each specimen, and sold three thousand copies costing one shilling each; the plant engravings for the plates were made from paintings done *in situ*. Though Curtis was more intrigued with distant and exotic flora, he wasn't adamant. Flowers familiar to English gardeners were included, many drawn from his own garden at Lambeth Marsh. Editorial policy was pliable as tastes varied; in the Victorian era orchids dominated the illustrations, to be followed later by rhododendrons. Here, too, was an outlet for the numerous discoveries of plant collectors such as William Lobb and David Douglas. Not until after the Second World War did hand-colouring of the plates actually cease; and in 1984, after a fairly fraught history, Curtis' *Botanical Magazine* was incorporated into *The Kew Magazine*. Ray Desmond, in an article entitled 'Two Hundred Years of the Botanical Magazine', which appeared in the July 1987 issue of *The Garden*, wrote: 'William Curtis' creation is invaluable not only as a repository of British botanical art but also as a mirror of gardening taste and progress over two centuries.'

There is no need to botanize or even, for that matter, to leave an urban surrounding to experience the sheer delight of botanical illustrations. This

floral art is certainly another whole world of gardening pleasure, far removed from the bent back and the earthy thumb. Museums, books and magazines tumble out a mass of glorious pictures, and whether you are a true botanist seeking information, or merely an uninformed flower freak, the genius and refinement of these plant illustrations spring from the pages. And the genre still goes on.

Margaret Mee, that indefatigable botanical artist, produced work of consummate exactitude. The result of those tough and tenacious journeys through the forests of Amazonia has bequeathed for posterity a unique archive. Many of the plants that she painted from actually studying them in their surroundings, are already lost forever in the perfidious wholesale destruction of great tracts of natural wilderness. That this remarkable woman should have survived thirty years of living amongst the Brazilians, studying their ways, their animals and plants, to have then been killed in a car crash in England in November 1988, only adds poignancy to the matchless legacy she has left us.

Looking at the work of contemporary botanical illustrators such as Margaret Mee, or the superb paintings of Margaret Stones, realigns our perspectives. Walking round the garden can be poles apart from just strolling and smelling the evening air. Examining the pared austerity of these botanical plates and studying the copiously rich bouquets painted by Jan van Huysum or Rachel Ruysch, make me think that staring hard at a crocus could become a singular pastime.

At the heart of gardening there may need to be a belief in the miraculous. We look for the evidence of this when a seed in the earth transforms into colour, or a cotyledon transpires into scent, yet alongside this aspect of gardening is a prosaic awareness that the kind of garden we create must fit the landscape.

When you first look at a garden, wherever it is, or whatever its dimension, almost involuntarily you pigeon-hole it. The layout may be formal or wild; rustic, classical or a place to hang the washing, but however you identify it each one portrays an individuality, reflecting back the owner as well as reflecting its classification.

Gardens should retain their personalities. Even though gardeners may copy bits from places they have visited, they should still use these transplants in such a personal way as to fit with ease into the setting in which they are created. Tunnel vision may work well as you walk from plant to plant, but at some perimeter the garden should integrate with what lies beyond.

Nowhere is this more critical than with gardens in the countryside. In the eighteenth century artists were so influential with their paintings of undulating pastures and bosky hollows that they inspired gardens which melted seamlessly into the horizon. That was until the advent of 'Capability' Brown who destroyed so many ravishing places in favour of grand sweeps of lawn, with bunches of trees congregated here and there as in public parks. But our pastoral gardens should not become parks in miniature. Kew Gardens and Central Park, Tokyo's Jindai Botanical Park or those in Sydney are places of curiosity with their vivid plants and elaborate beds, but why carry this image of urbanity into the countryside? Without realizing it, are we trying too hard for pictorial perfection jammed with cleverness and an insatiable compulsion for every new flower on the market? Have we become too set on creating an immaculate arrangement clean as a whistle in the middle of meadows?

It is this preference for order and control which now, on a much smaller scale, has invaded our own contemporary gardens. Walk through a village of a few hundred houses set down in a pastoral background, and garden after garden will conform to a blueprint of utter over-nicety. Neatness is as in-built as the prefabricated sun lounge and the obligatory shorn clump of violas.

William Robinson, in spite of calling one of his books *The Virgin's Bower* – which might make that part of the garden a bit underpopulated – does sound fairly dyspeptic at times. Writing more than a hundred years ago most trenchantly on gardeners and their gardens, he works himself up on the subject of unity between house, garden and landscape: '. . . and the best way of effecting that union artistically should interest men more and more as our cities grow larger and our lovely English landscape shrinks back from them. We have never yet got from the garden and home landscape half the beauty which we might get by abolishing the patterns which disfigure so many gardens.'

He carries on delightfully for pages on his prejudices; his testy comments are relevant to our own gardens of today which would, no doubt, make him blow his top. Here he goes again: 'The really artistic way is to have no preconceived idea of any style, but in all cases to be led by the ground itself and by the many things upon it. Why should we in the plains or gentle meadows of England not give effect to the beautiful lines of the landscape,

and make our gardens harmonise with them?'

Why not indeed? There is still time to ruffle up the over-contrived disasters we have achieved; to realign our garden productions and think again. Though the contemporary preference for mowing beyond the front gate means genocide to buttercups, it is not too late to loosen our control and allow a modicum of floral anarchy.

Coming nearer in time, there is that fascinating book which surely must have already become a classic, Jane Brown's *Gardens of a Golden Afternoon* – on the partnership between Gertrude Jekyll and Edwin Lutyens. Here there is an account of Hestercombe where Lutyens used textures of stonework for pool edges, pillars and paving and 'for the gigantic walls which relate the garden to its surrounding landscape'.

I know I go on about this, but when a book with the unabrasive title *The Country Garden* came out last year I felt incited to respond. What John Brookes with quiet wisdom was telling us was incontrovertible. His point of view goaded me to write:

> The pollution of the countryside is not only a matter of nitrates and silage effluent, of urban sprawl and brightly-lit petrol stations; a far more subtle contamination comes from tidiness. The obsessive and contagious mania for suburban conformity reaches out into rural habitats. Mown verges, white chains and swinging name-boards are seeping down the lanes. Plants acquired by the car-load from the invidious garden-centre, are placed with the precision of pieces on a chessboard.
>
> What people seek in the countryside, for its serenity and its unchanging visual allure among the innocence of cow parsley, is being eroded before our eyes. And it is done by the very people who have come to live amongst it. What has gone wrong? At what point does the motivation to live in unspoilt country become a compulsion to turn frenetic park keeper?
>
> Something must be done to stop the dead hand of banality from placing hanging baskets or dwarf conifers on the threshold of village cottages. 'Suburban' gardens are heavenly in the suburbs. That order, those cherished back gardens, the gaudy bedding plants and jazzy cherries all blaze away in their right environment.

The right environment is the point. John Brookes was reminding us that country gardens are surrounded by country; that a sensitivity to the locality is crucial to making a garden which sympathetically lies down into its setting. If we are not to be found guilty in twenty-five years' time of indolence and supine unconcern, some jolt is necessary to stop what is happening. A look over the hedge in good time would be providential before planting a Chusan Palm. Now I won't say another word.

Ideas for planting our garden gestated slowly. We didn't want to impose a plan from above on to its natural shape. Rather we tried to imagine the garden from below, where shrubs would seem to erupt naturally through the grass as though nothing had been contrived. It may sound idyllic and parts of it are, yet still we seem to be moving things planted in the wrong place year after year: trees, ground cover, honeysuckles or roses.

Roses. It's a word so integral to gardens it barely stops the eye on the page. But think for a moment just what roses are about. Our minds fill with faded childhood memories. Not visual evocations so much as a response to the word:

> . . . I watched in vain
> To see the mast burst open with a rose,
> And the whole deck put on its leaves again.

Who was the stranger who named 'A rose-red city – half as old as Time'? Or what child hasn't thought of the Wars of the Roses as a wild battle of red and white flowers hurled by infuriated warriors? There are so many associations, each of us must have a trail of rosy recollections. I know that winters can be spent absorbed in rose history, their nomenclature, ancestry and literature, when image upon image passes like a mobile palimpsest in the mind's eye: symbol of love and virtue, the Virgin Mary, England, York and Lancaster; distant origins from China, India and Persia; Minoan jewellery, Herodotus or the crusaders and the stained glass of Chartres.

> Oh, no man knows
> Through what wild centuries
> Roves back the rose.

Michael and I knew nothing. Instead, when David Austin's roses were in full spate, we visited his nursery and made notes.

But here lay trouble. The names of these wild and old shrub roses are evocative, all plushy romance and innocence: 'Reine des Violettes', 'L'Impératrice Joséphine', 'Robert le Diable' and 'Hebe's Lip'. Looking, touching and smelling is how we made up our list of first plantings. What we didn't know, for no book had forewarned us, was the will of the rose. It's an attribute of roses that is so crucial. Had we known beforehand we would not still need to spend autumns evicting a well-rooted plant from one place to another. Each rose has its own characteristic way of filling space. Each rose has an ineluctable habit, from the supple grace of 'Ispahan' to the glorious but inflexible 'Blairi No. 2'. This rose, for instance, sumptuous and scented, is so thorny and unrelenting that nothing can be done to persuade it to hang in sprawling festoons across an arch, which is what we had intended. Whereas 'Belle Amour', with her coral-cupped petals, is a spicy and ancient Alba who would have obliged with courtesy. So would the rambling 'François Juranville', with apple-scented, muddled petals. Roses can be assertive and spirited, or they can droop and languish; others are just groggy.

The implacability of the rose is something we have come to acknowledge since our first impetuous plantings brought back from David Austin's nursery. It isn't that we bought the wrong roses, it is just that we put them in the wrong places. The intransigence of roses is something we have had to accept and now, amongst our eighty or ninety different kinds, I know which are docile and benevolent from those which are headstrong; those that are pliant from those which are pig-headed. If we had known the mannerism of 'Celestial', with her fine pointed buds and fragile petals, clear pink and sweet-smelling, we would have made use of her lax capitulation and planted her over a low wall. Even 'Stanwell Perpetual', though immensely prickly, is willing for its arching branches to be manipulated. But not 'Maigold': spiky and energetic, with glossy leaves and superb coppery flowers, it needs space to show off its magnificence.

A couple of Centifolias, 'Fantin Latour', so shell-pink and sweet any painter would be glad to share his name with her, and 'Chapeau de Napoléon', which smells of a curious blending of lemon and lavender and has an intriguing green-tufted calyx, are two of our roses which I cherish with passion. It doesn't always seem to do them much good as some years they aren't as bountiful as in my mind's eye.

Then a Damask of such superlative beauty I fuss endlessly wondering if it's in the right place, is 'Quatre Saisons' or *Rosa damascena bifera*. It is cupped and quartered with crumpled petals, a very ancient rose surviving down the centuries with an elusive frailty which is at the heart of roses. Two Bourbons, ravishing and irresistible, are 'La Reine Victoria', lushly pink with layers of scented petals, and 'Louise Odier'. This is a rose of such adorable refinement you can't imagine her slightly lilac pink, deeply cupped petals being arranged in any other way.

Roses with a compliant disposition are an absolute godsend to a gardener. They fuse and merge with a natural instinctiveness with under-plantings of huge mauve or purple orb-headed alliums, pink and white verbascums and campanulas. These trails of blue flowers twine amongst the velvety-purple petals of 'Cardinal de Richelieu' or the Portland rose, 'Comte de Chambord', small in stature but potent, with quartered flat flowers of pink turning lilac in old age.

How easy to be carried away at this time of the year. It is the crescendo of summer when almost the whole garden moves too fast. Things are ripening with copious extravagance before our eyes: buds and petals, pinks and whites, foliage and smells. And all so ephemeral. A whole year's gratification can be lost if limited evenings of wandering from rose to rose aren't relished to the full. How imperative, therefore, is the leisurely evening ritual with secateurs, giving a reason for hesitating at each bush to appreciate the meridian of a year's work. Even thinking of them now, in mid-winter, I can conjure up those early mornings when it's dewy and still; when the bright green-leafed Rugosas are hummocks melting with colour. The fact that their summit of blooming lasts for a mere three or four weeks adds to the intensity of pleasure. If roses flowered throughout the year, think what we'd miss. That wonder in winter of putting out a hand to those dreary sticks and knowing that somewhere is hidden the culmination of next summer.

As the season dwindles and the last of the flowers leave nothing but reflected pools of petals on the ground to show where they've once been, I go round to do the final dead-heading. Not to all the roses; some I leave as feral creatures untended in the wild; some I leave for their flamboyant hips which hang about the boughs like polished cherries. Dead-heading is a job to be done in the evening, when the shadows are long, when a kind of serenity suffuses the garden, so different from the freshness of early morning rose gathering.

Not understanding the character of roses was one of our mistakes; there were others. 'Guinée', who has a kind of indoor smell of old dressing-up

clothes, is the darkest, deepest crimson climber, whose petals have a bloom like velour. She has been a total failure for us, though we have tried various aspects for her resplendency. So, too, has 'Souvenir de la Malmaison'. With a name like that, with all the associations of Redouté's *Les Roses*, we had to have her. Yet in spite of her delicious scent of peaches and her flat blooms, every year her buds turned into squashy brown pâté long before they had shown us their flesh-pink petals. After six unproductive summers we have banished her. 'Mermaid' succumbed to unhappiness, overcome by incompatibility she never made it though we did try her against various walls.

Most of the roses are resilient. They have withstood our climate with far more perseverance than many other plants, putting up with long wet summers that roses so abhor. The worst element is the wind. In winter the north-west wind tears at the roses on the corner of the house with relentless brutality. Even twelve-foot 'Mme Grégoire Staechelin', established for six years and displaying uninhibited magnificence, was struck down by a piercing winter. The wind and ice that year were so devastating that creamy 'Félicité et Perpétue' was annihilated and sadly it was also curtains for 'Emily Gray'.

If you garden you think about gardens. Ideas keep manifesting themselves, they seep into your mind often when you are nowhere near a garden. At a bus-stop or in the bath, potent thoughts erupt, having been triggered by some far-off glimpse, a description, a suggestion or a painting. Colours can fill the imagination, impossible to achieve but intense and real all the same. The spectrum of the inner eye is vivid and enticing.

One of my follies was brought on by a few drifting thoughts at some inappropriate place. From seeing roses falling down the canvas in Florentine paintings came an idea which later I discovered wouldn't work. Yet somehow that notion of single blooms against the sky miraculously descending from nowhere was tantalizing – as though there had been a cloud-burst of flowers a moment before. Perhaps this effect could be achieved with three standard rose trees carefully placed on the back bank so that coming out of the house the dangling flowers would be seen against the sky. Grey or cerulean, the colour of the sky didn't matter, what did was the illusion of airy descent, of unearthly apparitions suspended in space.

The production began inauspiciously. Firstly, a five-foot weeping standard requires a strong support. That meant an ugly stake with tough ties to hold the rose in a firm grasp. Then strong winds caused chafing where the roses had been budded, so lumpy padding was needed; next, those lovely lithe flowing branches had a bad tendency to turn up at the ends. Only by hanging stones from the tips of the branches could I get the shape I wanted. Already something was happening to my alluring vision; it was beginning to come adrift.

However there were moments, just a few, when we did walk out of the back door and saw roses high against the sky, 'Albertine', 'New Dawn', 'Félicité et Perpétue', and they did look the way I had imagined them. 'But what about the rest of the year?' any sane gardener would ask. How pertinently right they would be. For the rest of the year we are left with the stout stakes bound to rose trees with drooping branches held in shape by stones on the end of strings. Much as I love stones in a garden this arrangement was not exactly a work of art. The idea had seemed a good one for those few spellbound seconds when it first entered my head, but its realization was a botch. Fortunately I have been saved any further decisions because two days ago one of the three roses collapsed completely under the weight of snow, so that has scuppered the whole enterprise quite succinctly.

I blame these fantasies on those isolated moments when, undemanded, garden ideas germinate. I see I should have kept my head, but a part of gardening must surely have come from losing it? Without being led astray from the known and tried, how would Charles Bridgeman have conceived the idea for the first ha-ha in 1712? Vita Sackville-West contrived a clematis 'table' so that she could gaze lovingly into the upturned faces of the flowers; and wasn't it Gertrude Jekyll who first thought of growing ramblers horizontally as ground cover? Lady Anne Tree has a dressing table of yew, a four-poster bed made of clipped box with a vine canopy, a bedside table of ivies and an armchair of briar roses. As for outlandish garden eccentricities, they burgeoned from the dotty nineteenth-century Frenchman Audot, who made whimsical fantasies from sculptured trees, and his batty compatriot the conductor Louis Antoine Jullien, who cut his evergreens in such a way that a howling gale played the opening bars of a Beethoven symphony, to the giant shell in which to bask at Strawberry Hill, and the invention of glass cucumber straighteners. Thank God there's no limit to fanciful garden deviants. So many delightful whims and caprices lie on the other side of *Gardeners' Question Time* on the BBC.

Before leaving the compelling subject of roses, I should explain how our roses work. Our plan to grow them out of grass rather than from well-worked-over flower-beds, has been a success. It is not infallible, but we have found that mulching in spring does wonders, for it retains moisture and keeps down the rank grasses, preventing them from growing through the shrubs. We soak wodges of newspaper in a bucket, spread them round the base of the rose, making a circumference as large as we think the rose will eventually take up, and then cover it all over with two feet of mulch. This may be made from sodden hay unfit for animals or from fids of last summer's grass too coarse for compost. For special roses we use compost, rotted manure or leaf mould.

Mulching roses is a placid occupation, carting the stuff in a barrow from bush to bush, but there are some roses which are devilish. *Rosa paulii* is infernal. It requires immense will-power to get Michael and myself out on a bank where three roses rising like foothills need to be mulched in spring. Protected by leather gloves, tough jackets, knee-length boots and hats jammed down over our ears one of us has to use a pitch fork to hold back the trailing boughs of this brutally thorny rose with titanic strength, while the other goes into the bush bottom first with armfuls of wet newspaper and mouldering hay. It is not an occupation to bring out the best in you. If we are speaking at the end of the afternoon it's only to gasp, 'This must be the last time.' It never is though, because come summer that whole bank is a waterfall of trailing tendrils covered with bone-white flowers smelling of cloves. Memories of the indescribable afternoon spent mulching are misted over; we know that a celebration of *Rosa paulii* has to be an annual event.

Roses designated as non-desirable because they don't do well indoors carry no weight with us. The most dispiriting words in a plant catalogue are those which advise: 'Ideal as cut flowers', or 'Popular with flower arrangers'. These recommendations must refer to those tight-lipped affairs which have no affinity with the roses that grow in our garden. In fact when I read of this classification beside any flower in a catalogue I instantly condemn it, for by implication it doesn't sound like much of an asset for a garden. It's sad about flower arranging anyway. It has become far too precious a pastime; too consciously artistic and contrived. The way a child gathers flowers and puts them in a jar has spontaneous charm, completely lacking in those over-elaborate and sometimes quite hideous confections learnt from flower-arranging classes. Yet a series of different coloured glass bottles of varying size each holding one flower, form a composition on a window sill where the light illuminates the petals and lights up the glassy iridescence whether from transparent, blue or green bottles. And should you want to, this method also has the merit of allowing you to cast a sort of Margaret Mee eye at pedicels and internodes as well as corollas.

Summer is the competitive season. It is when we set ourselves such high targets; when failure is a built-in contingency. Summer days are days of obligation, when things are needing us for water or for admiration. Palmy days start us off from almost six in the morning with their dewy flowers, an infinity of smells and all the compelling demands for our boundless appreciation. Oh, it should be good. It is. But there is that quality of breathlessness when we are running hopelessly to keep up with the turbulence of summer. We need our perceptions at the ready – we mustn't miss a thing. Now is the summit of our endeavours – the climax and justification of hours of toil. So I find myself competing to keep up, knowing that though the bountiful days last till ten at night yet still there isn't enough time. It is sheer, blissful hell.

Then there are all the decisions to be made. For this is the season for sitting in the garden – but where shall it be? Choice is impossible, everywhere is perfection. Under a tree for dappled light? By the stream for its calming sound? Beside a rose drowsing in its own scent? But which rose? Shall it be the grapefruit-scented 'Mme Isaac Pereire' or the pure, clear fragrant *Alba semi-plena*? Sitting in gardens doesn't seem to be a built-in requisite with gardeners:

> Our England is a garden, and such gardens are not made,
> By singing – 'Oh, how beautiful!' and sitting in the shade.

How stuffy of Kipling. How priggish, and anyway, why not? Surely ruminating and lolling, squandering slivers of time as you ponder on this or that plant; perching about the place on seats chosen for their essential and individual quality, are other whole aspects of being a gardener? Why shouldn't we? We sit in other people's gardens, why not in our own?

I know the answer of course. Only phoney gardeners sit, real gardeners can't. In no time at all their eyes are fidgeting from one error to another, they long to be pulling up a stray weed, tying a drooping stem, or are just left seething with frustration that every moment spent immobile means one less spent on doing essential chores. Michael and I work differently. It may be that having spent seven years in Thailand, where it was far too hot to do the gardening ourselves, we have grown accustomed to sloth. Sitting in

the shade, amongst the hibiscus and bougainvillaea, seemed a perfectly reasonable and most civilized way of behaving in a garden. In that country our sitting always had to be out of the sun, but here, in England, the scope for subtle sitting is infinite.

We have places chosen for each view, smell or time of day. Up the orchard under the boughs of the old pear tree for an unexpected warm day in May, while the blossom is on the trees, or on a little stone 'shelf' literally dug into the stream bank so that with my feet almost in the water, I can sit excluded from outside sounds because the rippling of the brook is dominant. We have a bench under a guelder rose within easy reach of the house, and a secluded seat amongst the fourteen hazels we planted seven years ago and which are slowly forming an irregular circle of total greenery. Another bench is placed for a long view up the brook, another is of stone, low, against a wall in the midst of all the herbs giving almost a bee's view of what it must be like to spend your days head-first in borage. One of the most satisfactory places is from the high bank across the stream where we look back at our garden seeing everything spread and multiply into the distance; or a place just for evenings where the disappearing sun still reaches to light up the flame-coloured flowers of *Lonicera* × *tellmanniana* and a shrubby plant, *Holodiscus*, with its small waxen flowers in long panicles.

But the best of all our seats is the one we had built last year. I wanted a bird's eye outlook on the garden and having trees growing down the stream, what better vantage point could be found than among them? There was only one person we knew who wouldn't be daunted by the prospect of constructing a tree house for such a position.

Richard Craven, who lives in the next village and has made a lych-gate with a fish-shaped latch, garden seats that integrate into their setting and bowers and casements, didn't hesitate when asked to make a platform and seating for two up a double flight of steps. A more cautious person would have boggled at the practicalities, but Richard seized on them as trivia. The problem was our trees. None were sturdy enough to support the structure, so Richard planned it to be attached between two alders. Trees grow and trees as slender as these sway in the wind, which made it imperative to have a flexibility to the affair. We left the problem with him.

Some time later, on a grey day in March, Richard arrived with the tree house constructed. By means of pulleys we managed to haul the contraption up between the trees where it was secured by hinged bolts. Except for the seats and handrail, which are of elm, the whole thing is made of oak, cleft oak, so that there is a natural curved movement and grace to its design. The treads of the steps, the handrail, the camber on the seats and the concave rails have a flowing movement, a kind of living continuity with the trees themselves. Like the ark everything is held together by wood, and mortice and tenon. In a few years the untreated wood should weather to silver so that in winter, when the leaves are off the trees, it won't stand out as something eccentric. It doesn't; we can sit there on summer days amongst the leaves and look down on the brook fifteen feet below us, or out between the branches for an aerial glimpse of roses.

Richard devised one further refinement: a circular wooden tray, specially designed to hold a bottle and glasses, which can be hauled up by a rope over a wooden pulley set high in the branches above the platform. While the bottle is cooling in the stream, we can sit secluded and unperturbed. On some days the leafiness is static, the atmosphere soporific; on others the seat sways as imperceptibly as breathing, while the surrounding leaves produce a continuous sibilant lisp.

Summer is also the season when some of the hideous flowers are on display, flowers which are uncooperative, or which eclipse others with their blatancy. Harmony may include the sharp pink of godetias or the acid yellow of mulleins, but so much depends on how they are used. At Pusey House in Oxfordshire, there is a herbaceous border of such unsurpassed splendour that only by taking a long look can you begin to unravel just what has made it so beautiful. Verbascums, delphiniums, foxgloves, heucheras, achilleas and phlomis in generous blocks of colour.

Not all flowers are things of beauty, in themselves or put with others. For example dahlias aren't easy on the eye and, personally, I find *Hypericum calycinum*, 'Rose of Sharon', a most joyless thing, used like blotting-paper to soak up space. Then heather! How it mutilates gardens with its puréed fruit-pulp appearance, its neutered growth and depressing meanness. Faced with variegated plants I feel dejected. They have an aspect of malady, of something not quite right. I don't mind that unnatural albino look on leaves of apple mint, but dogwood, *Cornus alba* 'Elegantissima', variegated grasses or euonymus have a loss of pigment characteristic of some congenital malingerer.

Effete lupins or doleful bulks of rhododendrons do nothing to make my hair stand on end; neither do raucous gladioli which might work growing on the back seat of a sports car, but which look spitefully aggressive lifting their

heads above the pallid hues of docile flowers. *Leycesteria formosa*, with its dangling racemes like purple-black ear-rings of funeral jewellery, are murky tassels of little distinction. I know I ought to admire *Tropaeolum speciosum* because so many good gardeners take trouble to grow this twiner with pretty vermilion flowers, but straying over a yew hedge it looks like a thread of cotton needing to be picked off. How much better if its startling little flowers were seen tangled amongst the purple flowers of a clematis, or climbing up the dark grey of a stone wall. Cotinus are hard to place; we haven't got one but I look at other people's. Some gardeners have a genius for fitting that bush, the colour of broken veins, into a suitable background, others leave it abandoned like a great beached summer pudding in the middle of the lawn.

Yet even as I write this I know that these are small complaints. In summer gardens generally there is little to disparage. Our planting mistakes in this garden are unending, but in the evening light of late July, when the shrub roses are past their best, I can walk round the pond and see yellow candelabra primulas, creamy meadow-sweet and tall fading spires of ligularias reflected in its murky green waters. All is so still. Against the wall the blooms of 'Mme Alfred Carrière' are drooping from the weight of their own opulence. Yellow flowers of the rose 'Leverkusen' almost shimmer, thrusting outwards in that rigid habit of each flowering stem. And still visible in the waning light is the white effervescence of newly opened 'Longicuspis', spilling its small flowers over a wall and among the leaves of a May tree. This is almost the last of our summer radiance, these fading yellows, creams and whites pouring out a final abundance of trailing blossoms and falling petals. This is also the time to stop.

Is it possible to garden and to travel or are the two incompatible? If you talk to gardeners they always seem to be waiting. Anticipating and hoping; watching in anguish or jubilation, forever urged on with greater ambitions. Can gardeners be otherwise? Can there be a kind of half-gardening or must commitment be total? If gardening is the essence of continuation, of one thing flowing into another, a procession of followers that must be compounded by colour, habit and form, then where does that leave travel? Almost in total opposition. An impossibility; if you leave a garden for five days disaster may occur which no amount of catching-up can later eradicate.

Yet a garden and travel is what Michael and I wanted. A time for cherishing and a time for wandering; and the only way was not to be ambitious. Not to want too much happening in the garden all the time, event after event. And most important, not to regret owning a green garden. It's hard, but there is little alternative if you know that the recurring travelling fever is going to break out in a month or two.

Each May there is a pause, or we say there is, though we know the garden is enticingly on the move and every bud compelling. But we say there is a pause, just long enough for a week or ten days' travel. The wrench of laying down the trowel and turning our backs on all that early summer fertility is not as harrowing as anticipating it had been. For once we are away, headed for other countries, other smells, the garden remains in my mind suspended, inert and unforthcoming, merely waiting for our return. Greenfly, drought and moles are phantoms of someone else's nightmare.

On returning the garden is then paramount, we are moving inexorably into the season of peonies, wistaria and roses. The garden unequivocally dominates our lives, our commitment is total. We don't even open an atlas. But once the roses are over we have to be indomitable. No wavering or longing for just a few herbaceous flowers, a compromise of beds, a weakness for an armful of delphiniums here and there (for how I love them). We look at other people's gardens but we mustn't hanker; we enjoy their lovely summer-flowering borders, their dense beds of colour and all the scents, but we mustn't yearn. As our roses fade, so must our acquisitive instinct for prolonging the summer season. It's not easy. It requires stubborn single-mindedness. One exquisite *Eucryphia* × *nymansensis* planted in a moment of wanton exuberance – and we are hooked. How can any gardener turn away from buds in their first year's flowering?

There is one bonus, omnipotent and cardinal, for travelling gardeners, a compensation that should not be underrated, and that is coming home. If you never leave your garden you can never relish that frantic impatience to get back; to walk round together assessing with delight or despair what is thriving, what is wilting. When we have satiated our travelling appetite, there is always this one further gardening anticipation which is very real, very constant and very potent. The better the garden, the more compulsive the reason for leaving it – the heady delirium of coming home is in exact proportion.

5 Bulbs, Corms, Rhizomes and Such

WHY GARDEN? GOD KNOWS. The onus, guilt and compulsion hang round my half-sleeping thoughts in the autumn days when the great bulk of summer has dropped over the horizon, when the clocks have changed and as a dormant animal, I'm looking for hibernation. Damn those fine mornings. It's then the guilt seeps in like a bad gas. All the things I've been saving up in summer to do later, comforting myself that the season will change and the garden won't ask for my whole-hearted committal, my unalleviated devotion, when really I shall have a chance to consider things other than that floral blackmail outside. And what happens? Why in autumn, people and books tell me I should be out there making the best of every opportunity to be preparing for winter. I hoped it could be left. The garden I mean. I thought all that messy growth of shrivelled leaves and crispy stems could remain as protection for and comfort to the plants against the frosts to come. I know for instance that the peonies, naturalized in grass, have lasted countless years because their useful canopy of decaying leaves protects those tender fat crimson buds which appear astonishingly early in the year. But it isn't like that. Real gardeners will say with a glittering eye and a surfeit of energy on some golden October or sombre November day, 'Isn't this perfect for being out in the garden?' Is it? I haven't the faintest idea what I should be doing out there. I don't want to know. For myself I'm already facing the other way; my sights are fixed on everything splendid there is to be done that has nothing whatsoever to do with gardens. The garden should be sighing and settling itself unaided into contented slumber. It is the season of sleep, of torpor, of a lack of sap and fecundity. It doesn't need me, surely?

That's what I do like about gardening. The seasons. How blessed we are to have them. Imagine living where there was only a monsoon and a hot season, or where snow lay metres deep for months; where spring lasted a fleeting few days, and where autumn did not go gentle into that good night. How lucky we are to have weeks with first the ashes turning gold and then, much later, the oaks. How fortunate to have the lingering days of autumn when imperceptibly things begin to wither; there's a time to gloat over ochre mounds of Rugosas with their fiery hips, when leaves on the vine turn russet and a viburnum has corroded to cinnabar; against the house the tentacles of hydrangea are leached to lemon yellow.

The tulip tree, for the first time in the year, is dominant; its incised leaves turn aureate in sunlight while the milky petals of a shrubby hydrangea become speckled with pink and the colour of coral seeps into the green of its leaves. My spirits rise at this autumnal decline.

On some winter day when the sky and the landscape are achromatic I find the garden at its best. Everything is latent; there is an undertow to the garden and I sense that below my feet is the whole of summer. Not the one that is past but the summer to come; when as yet there is no need to face disenchantment.

When the frost lies even across the nearby uncut summer fronds, across the distant meadows rising from our garden to where the sheep and cattle stand motionless, their breath carving the air with evaporating shapes, involuntarily I'm drawn outside. The stillness is so immutable; the calm season so absolute there's no need to feel impatience for spring. Walking over the crunching frost with aching toes, relishing the bitter grip on woolly fingers and blowing them back to life, I fill my lungs with air like crystals and wonder that these rigid stems, black but traced with rime, can mutate into voluptuous domes of roses. Summer amplitude, barely recollected, belongs to some other continent; it's now that counts, when the garden is deep in hibernating peace. The idea of 'making the most of the weather and getting things done' is an irascible thought.

We never manage to have things blooming all through winter as true gardeners do. They always have something for colour, a few blooms or

stems, a few fragrant shrubs in the midst of frigidity. Even so our garden compensates for my laziness and lack of endeavour by making a walk in winter a series of discoveries.

How contrary plants are. How capricious. Some, like lavender and herb robert, look perky on a frosty morning, whereas bergenias temporarily turn up their toes, their fleshy leaves lying limp and hopeless. Out of walls along the brook appear the shiny leaves of hart's tongue, glossy and bright, bursting with good health in the freezing air. Wintersweet, in spite of its name, has knuckled under yet the roses 'Albertine' and 'Mme Alfred Carrière' are impervious, their leaves don't even acknowledge a change of temperature.

The ground fragments under my feet, my breath steams, and I am not sanguine when I see the beauty of santolinas extravagantly frosted on their silver leaves; last year they succumbed completely. At the foot of a tall, leathery *Clematis armandii*, here for the first and no doubt last winter if this keeps up, are some campanulas in bloom and a wild strawberry has one white flower, yet nearby cranesbill perversely sags with cold. I like the way plants such as Canterbury bells, London Pride or primulas crouch low to the ground in frosty weather; they flatten themselves like ears on a soppy dog trying to ingratiate themselves to this piercing inclemency.

Not only are some plants insensate to the bitter cold, but some winter morning we'll hear the dipper warbling its melodious ripple of notes from somewhere down the stream, for all the world as though spring were here. Occasionally it perches on a large earthenware pot beside the pond, cocking its head at the frozen water where deep down is a dim luminous glow from the fish motionless under ice.

Later, when somewhere in the world glaciers are beginning to creak, in the garden the earth is already heaving with unseen life; snowdrops and aconites appear. Spring unfolds in spasms and setbacks; crocuses, scillas and daffodils emerge and sometime there will come one unbelievable day when the warmth and fertility are genuine. When pussy-willows and the balsam poplar fill the blue sky with their outlines and scents. When the species tulips, the *turkestanicas* and *tardas*, draw us back throughout the day just to marvel that these little honey-coloured things with their yellow centres belong to the same family as their purple and scarlet regimented relations in St. James's Park. And later, after the early species tulips, there are others to follow which are utterly different. They too come up in the meadow grass, but unlike those earlier ones with their undemonstrative behaviour, these are the taller, more majestic lily-flowering tulips: 'Ballade', 'Mariette', 'Jacqueline' or 'Lilac Time'. These defy even our most optimistic belief and ambition. They really do surpass our maddest aspirations, for they go on blooming through six weeks of our prolonged spring.

Lily-flowering tulips, with their reflexed petals, grow to a height of about twenty-two inches. When the sun is low we can see these hundreds of translucent colours, pinks, mauves, crimsons and vermilions with the light shining through their curling petals like colour through a stained glass window. As the weeks go by, still every morning we can walk down the orchard to see them, the grass has grown up to their height, and the blurring and diffusion give another effect, that of an impressionist painting of flickering stippled colour growing amongst the Sheep's fescue and Meadow Barley grasses.

Tulips certainly went to our heads. We wanted more. Any bulb catalogue sends you wild with its delectable descriptions detonated across the pages. There is almost no corner in a garden for which the perfect tulip for that particular situation does not exist. Last year we lopped off a piece of our six-acre field so that now the land flows down into the garden, or rather that our garden continues seamless into the countryside. Here our longing for more tulips can be fulfilled. We've planted hundreds of small double early bulbs: white and silver 'Schoonoord', pink and white 'Peach Blossom' flushed with carmine, and 'Garanza', a volatile and explosive pink.

Bulb time is a season to be relished; a time to remember how fortunate we are not to have a hot summer that comes overnight and burns up this benison of English springs.

Certainly if I were starting again with a garden like a sheet of unmarked paper I would plan for bulbs. I know this is not a new discovery, unique and amazing, and I've been particularly slow to grasp the versatility of bulbs, their endless variety, their individuality and, more than anything, their scope for discreet charm, sparkling radiance or regal formality.

What makes bulbs so special? I think it has to do with grass. There is something ineffably touching to see a flower as slender and fragile as the snake's head fritillary, *Fritillaria meleagris*, pushing up through the tough thickets of unkempt grass last cut in autumn. Those undemonstrative dangling heads of mauvey chequered patterns with their important Hellenic names of 'Artemis', 'Charon' (appropriately coloured blackish-purple), 'Poseidon' and the chaste white 'Aphrodite', with a slight green luminosity

about her. I want them all. I want them to spread themselves extravagantly about the orchards.

In contrast grape hyacinths appear with vigorous tenacity thrusting up their bobbly heads in arbitrary clumps of textured blue. Snowdrops, chionodaxae and crocuses appear through tufted grass with comforting dependability. Tolerating even longer herbage is *Camassia esculenta*, Bear Grass, with its intensely blue starry flowers on tall stems. We have a small patch of this strewn under young willows, its determined heave through the long meadow growth of early June is another statement of the resilience of bulbs.

As for *Anemone blanda* I just go weak at the knees for them. Those wide blue faces opening to sunlight some unexpected March afternoon are one of the high spots of gardening. How can they do it? Where do they find the blue? We watch them for weeks, observing their intricate leaves lying close to the ground like the fingers of a pianist before he starts to play. Suddenly, one day from a room in the house, we catch sight of this improbable blue spread as cloth on a piece of ground. Looking down on their upturned faces it's hard not to believe that after this summer will be downhill all the way.

Dare we try lilies? When I read that the madonna lily, *Lilium candidum*, has been cultivated for three thousand five hundred years, I see I'm a bit slow on the uptake. But what a choice. How shall I decide from amongst all those hybrid beauties: Trumpet and Aurelian, Martagon, Longiflorum and Oriental? I long to have them but so far lack the nerve.

One of the recurring pleasures of planting bulbs in autumn is walking amongst them in spring and wondering how on earth you did it. Bulb planting is torture. The basket never seems to diminish and yet stamina in the lower back does. Eventually there seems nothing to pull you upright over and over again but a series of protesting muscles. It's one of the most laborious and tiring jobs, even so, lifting resistant clods of turf, carefully placing each plum-sized golden bulb in the ground, you know you are burying amongst the worms spoonfuls of colour: crimson, rose, purple or magenta; white flecked with mauve, pink running into amethyst and white. And finally, after straightening your back and stamping back the grass, leaving no evidence of your hours of labour, you can think: why, there's only six months to wait before being confounded.

Birds are tyrants. They have a tyrannous effect on gardeners. We long to have them with us, but to hear their polyphony of song at four on a summer morning is something so dominant that it's hard to turn over and go back to sleep. Outside the garden is fermenting; it's so early, there are no shadows yet, there's almost a bloom lying over the garden; dew glistens on leaves and the smell of fresh petals percolates the bedroom. That urgency outside is not always possible to resist.

The unavoidable commotion of birds is only one form of tyranny, another is their audacity. They will nest close to where I want to garden although they have a whole choice of places: discreet, isolated and tactful. Instead a blackbird nests at eye-level in the ceanothus. When she's sitting on her eggs I can't avoid that beady eye watching my every movement. She inhibits me; she forces me to crouch – to walk like Groucho Marx so as to keep below her eye-level. I don't feel like challenging her; I move rapidly, ignoring her presence because otherwise, for the rest of the morning I should feel guilty for having finally forced her off the nest. Instead, while she's sitting on her eggs, that bit of weeding has to be abandoned; that piece of garden is hers. After the eggs are hatched it's worse. Then both birds perch and rustle about in the weeping pear, their beaks stuffed full of insects, blackmailing me to get off their territory. If I turn my back and keep on gardening I'm only haunted by thoughts of those hideous, raw-looking young waiting hungrily to open their grotesque beaks for food I'm preventing them from having. In the end I move off, of course.

Elsewhere in the garden, where a tumbling climber is in drastic need of propping up against an old stone outhouse, I disturb a nesting fly-catcher. This discreet little bird has an unobtrusive way of making her nest in some tiny cranny and then, in spite of her diminutive size, exhuding a powerful sovereignty for yards around. How can I go poking about with secateurs and bits of string when each time I attempt to do so this tiny brown bird, with its swooping flight, will scold me impatiently with her grating 'sip-sip-see-sitti-see-see' from a nearby twig. Naturally I throw up the sponge and leave that corner of the garden until the young have flown.

In the end, in spring, we move like trespassers in the parts of the garden annexed by the birds; we cut holes in the garage for swallows to nest in, we walk by the other side of the pond when martins and wagtails are busily gathering nesting material in the yard; we circumnavigate an ivy-clad tree where a tree-creeper, looking like a piece of animated bark, has made her nest. There is no doubt that spring is the time for the predominance of bulbs – but birds, too, hold us in their thrall.

Whatever our garden is, it seems almost impossible not to yearn for something else. Whether your longing is for another country, another climate, or just to have a different altitude, view or soil. Those plagued by a windy site must spend hours imagining their perfect garden in a sheltered hollow; others facing sea spray, where their choice of plants is limited, might crave a garden landlocked by hills. Gardeners emblazoned with azaleas may pine endlessly for chalk soil and the filmy colours of lilac, the subdued blue of *Buddleia alternifolia*; those bent double over clay will bless their roses, but nightly dream of artemisia, whose silvery gray flourishes in sandy soil alongside the papery petals of cistus, thriving from such free-draining earth.

How traitorous I am to think with longing for a back garden. But I do. I still have not satiated my love of stone; the idea of a truly paved courtyard, say, where there wasn't a single chance of trespassing further than its boundary, fills my gardening greed with desire. How challenging and exciting to create illusion within the confined shape of something within walls, measuring ten yards by twenty-five. How intriguing to choose with discriminating precision flowers and shrubs constrained within limits, rather than as here, where we can go on bringing armfuls of land into our garden year after year.

As soon as I think of back gardens I think of front ones. You start with those; those lovely little areas on show. How pretty to see wisteria or jasmine climbing up between the windows, around the front door, and to see a few wallflowers or pansies in the limited space where things have to withstand the lethal breath of traffic. Walking down a street looking at them I can see we are meant to peer. They aren't for privacy but for adornment. On the outskirts of small country towns, where the front areas are larger, how flamboyant and dazzling some of those gardens are. In spring Michael and I have only to leave our cold hills here, where nothing is yet stirring, to be astonished every year to see crown imperials and flowering almonds, primulas and *clematis montana*; forsythia and flowering cherries push up the colour decibels even higher.

But at the back? Ah, that's another thing. There is something infinitely mysterious about back gardens. Though they are never private, especially those in cities where there is a terrace of high houses, yet there is that curious quality that lulls you into feeling secluded. Logic makes it obvious that we are each overlooked, but the strange thing is that sitting in a back garden gives you a sense of privacy. How absurd. Here in the midst of the countryside we would immediately notice people lounging in the field, or would be inquisitive if we saw someone basking on the bank. Yet it's not like that in a back garden, when sitting with friends over a meal there's a sense of exclusion: an unawareness of what's happening over the fence. Perhaps in street beyond street people cocooned within rectangles have created their own illusions of solitude.

No wonder those who garden for the first time in a city have an almost breathless astonishment at themselves for having contrived such places. They seem surprised at how well it can work. How what is usually associated with the country, with established gardens of antiquity and mellowness, can be cunningly modified into an urban retreat, where a little sylvan hoodwinking really works.

Back gardens can be almost sunless; facing the wrong way and with high buildings around and maybe with the whole area covered with concrete, it is hardly a conducive set-up to start a bowery of flowers. But we once saw a most ingenious scheme whereby the owner had overcome his sunless yard by flooding the whole place under a foot of water. Bamboos and water-loving exotics with diverse textures and leaf patterns had been planted in raised beds along the sides. In this dank jungle the boundaries were invisible, the neighbours obliterated. Large sawn logs standing in the water made it possible to walk in this small yard by stepping from log to log around a central island of royal ferns. Enjoyment of the garden was to be savoured from indoors, where the whole wall of the house was made of sliding glass panels. Looking out into that verdurous density we could well imagine hearing the 'chee-yoop' of a racket-tailed drongo, just out of sight in the swampy enclosure beyond the gunnera.

An essential element of back gardens is hiding the boundaries. Other illusions have been contrived by jutting walls, arches, hedges or legions of greenery that seem to go on forever. A back garden in the heart of Birmingham has just this sense of wilderness. In the middleground is an umbrageous yew tree rising from a haze of bluebells. The grass leads to distant hazels and guelders, sweet briars and hawthorn. Often a fox lies curled up for hours in a patch of sun fifteen feet from the house.

Why in back gardens do people not move away from the conventional plan of beds along the two walls and a lawn in the middle? David Hicks, in his book entitled *Garden Design*, writes, 'In a small town garden, it is usually a waste of time having grass, because it hardly justifies a good lawnmower, and is very time consuming: gravel, stone and brick can make an extremely pleasing terrace or central path. Small gardens need durable surfaces to withstand constant use.'

One garden we know where the owner has broken away from the lawn and border pattern, has its flowers in a massive, meandering block down the centre of the garden. Lilacs and other small trees, climbers and twiners such as *Schisandra chinensis*, with its sepals that flutter like myriads of pulsating facets of light and *Bignonia capreolata*, a dominant twiner with two-inch funnel-shaped flowers of a startling scarlet, dramatically hide the walls and obscure the next-door garden. Two narrow paths on either side of the central block join at the far end where there's a seat to look back at the garden. There isn't an attempt to create any space or perspective here. Instead the overwhelming effect is as if you had stepped into a basket of flowers. The paths are so narrow pollen brushes your hands and clothes; walking round, conversation takes place over your shoulder. The only concession to floral restraint is right outside the house where, grudgingly, kolkwitzia, lupins, escallonias, freesias, diascias, and delphiniums are restrained just enough to make room for a table and chairs. Even so plantain lilies and passion flowers may tickle your legs or droop over the salad. And here, too, there is one asset that often comes with town houses – on opening the front door you can look through the house to see a far glimpse of the garden.

Visual deception of a different sort has been geometrically devised in a long back garden. Strongly defined perspective is achieved by a series of half-walls at right angles along the entire length of the garden. Distance is emphasized by a statue on a pedestal. Within these sectioned rooms are disparate plants. On the shady side are ferns, maidenhair and pulmonarias, whose spotted leaves go on growing larger long after their flowers have died; also convallaria (lily of the valley) and climbers that tolerate shade. Amongst the many things loving the hot compartments opposite are scented herbs, alpines and bulbs, growing beneath a profuse overspill of wall plants garlanding from bay to bay.

Other spectacular effects, theatrical and inventive, were in a back garden with a high bridge spanning its width. A flight of steps led up to where a table and chairs allowed the owners to dine looking down on their flowers. Twiners wound round the struts and banisters; underneath, a painted doorway flanked by urns led nowhere.

There are also those back gardens that are tidy and fastidious for annual judging in local competitions. One impeccable creation after another is lovingly decked out with such superb, colourful variation they should be listed along with cheese and chocolate as potential hazards to migraine sufferers.

Each of these gardens is a miracle of escape: a haven of seclusion, of disguise and containment. I love the lot. Simple, purposeful or histrionic, their lure is irresistible. As I paddle among my stream-side primulas, survey lily-flowering tulips in the long grass or prowl among roses, I brood about back gardens. Here, amongst limitless space, I plan small courtyards or medieval sanctuaries: places of paving and pots, where an outstretched hand touches a definitive boundary. What bliss.

We are an astonishing island with our gardens; they are unlike anywhere else. Why? What makes us get down on our knees and tolerate so much exertion and dedication to make these ravishing places when we could be walking the hills or loafing in the sunshine? The French, the Germans, the Scandinavians don't go in for it in this way. They don't make these cherished retreats. Apart from the energy needed, think of the money we spend. One only has to look at a garden centre from March till November to see it thick with avid shoppers. Business in these places must be thriving; the nurseries are bursting with trolley-pushers seduced into buying more than they came in for. But why? I don't know; but given a space round the house it won't be long before the owners are out there with a spade making their own unique sanctuary.

There is no typical British garden really, though we do designate a certain abundant rosy look as characteristically ours. Bowers, pergolas, beds and borders support colours of effusive prodigality. Our structured formality of hedges, paths and walls; our topiary and paved steps have a certain mellow harmony quite distinctive of this country. If you fell from space into one of our gardens surely you wouldn't mistake it as being anywhere else?

Here are three gardens. Each one was started by its owners. They all come into the category of garden because there is no other word to define their range of differences. The first is in Scotland.

Imagine a garden of sixty acres standing on a wild headland reaching out to the Atlantic where things of such tenderness, such rareness, such unlikeliness are all flourishing in robust multitudes. This is Inverewe, a spectacular place now run by the National Trust of Scotland, created more than a hundred years ago on a remote corner of Wester Ross. But it isn't the wide path flanked by sea holly, woundwort and Spanish broom; neither the startling potentillas and cotoneasters nor the leathery, glaucous leaves of

New Zealand flax that quicken my interest, but it's the man behind the garden. He was Osgood Mackenzie, surely one of the outstanding figures in gardening history, where patience is a built-in prerequisite? With perseverance of monumental proportions, Osgood waited fifteen years while his sheltering belt of Corsican pines and Scots firs grew tall enough to protect his garden from the prevailing south-westerlies, blowing unimpeded from Labrador.

Now, if you walk there, maybe with the intention of marvelling at the bamboos, the magnolias or 'Himalayan Lilies', *Cardiocrinum giganteum*, reaching up to twelve feet in height, you should be aware of two things: firstly, that where there are now wide gravelled paths to allow over a hundred thousand visitors a year, there were once narrow grassy footpaths threading through ferns and rhododendrons; and secondly, that you must remember the man himself. Osgood Mackenzie, with the benevolent Gulf Stream lapping at his feet, had the vision and tenacity to grow exotic plants amongst the braes. Kindly thoughts and admiration may course through your mind as you look at a coral plant, *Berberidopsis corallina*, or at a lantern tree, *Crindodendron hookeranum*, both originating from Chile, but reading his book *A Hundred Years in the Highlands* the greater part of his account is taken up with slaughter. While waiting for his shelter belt to mature, in one year he shot nineteen hundred birds, including blue rock pigeons, and golden plovers, whimbrels, snipes and teals. Even his dog helped by digging out stormy petrels. Yet after steady years of carnage he writes plaintively in a chapter headed 'Vanishing Birds': 'Can anyone explain what has caused so many of our birds to disappear?'

The man is a paradox: in spite of his lust for slaughter his garden is a plantsman's paradise. At Inverewe he has made a place of imagination and beauty, and, after his death in 1924, his daughter Mairi – who had grown up to love the garden, knowing every corner and secret place – devoted herself to caring for it for the rest of her life.

Olearias, the daisy bush, grow like huge display creations so different from the tender little affairs that cringe from frost in parts of England; an evergreen, *Desfontainea*, with long waxy flowers of scarlet tipped with yellow, flourishes here and the meconopsis, with faces of such open candid blue, makes others look like ailing relations. In this lushly temperate world there's a tree of unsurpassed beauty, a *Eucryphia glutinosa*, and a flower of miraculous frailty, *Romneya coulteri*, fluttering its translucent petals like butterfly wings. Dazzling gentians, cistus and euphorbias, rhododendrons and eucalyptus all bask in the mild benison of this faraway terrain, where gales and sea spray can often molest the coast relentlessly.

There is something else at Inverewe besides the rare, the exotic or the bizarre – it is the perpetual awareness of the sea. Behind every sighting of tropical plants from America, Asia, Australia and Africa, just out of reach, sometimes out of view, is the sea. The sound and smell of it infiltrate wherever you go. From under some gargantuan rhododendron originating from Yunnan there may be a glimpse to the purpley distance of heather and highland colours on the far side of Loch Ewe. Or walking up the headland the astringent smell of wrack trails its briney scent, overwhelming the pungency of resin. High overhead the wind produces a continuous, mournful sound as it makes its way through Osgood's trees.

Visiting Inverewe I like to remember that dogged man with a trowel in one hand and a gun in the other, having the grit to make a garden from a desolate piece of moorland covered with stunted crowberry.

My second garden, Stone House Cottage Nurseries, is quite a different affair. 'We wanted to have six or seven different gardens within this one,' James Arbuthnott said. You might well reel at such a concept when you realize their garden consists of only three-quarters of an acre. You'd be wrong, however, unimaginative and stodgy, because six gardens are just what James and Louisa Arbuthnott have made.

Kidderminster in the Midlands, that are 'sodden and unkind', is an unexpected place to find plants that flourish in Ireland or on the west coast of Scotland, in Devon and in Cornwall. Three distinct features are dominant and at the very heart of what the garden is about. The first is the intricate design, meticulously worked out by James; the second, the wonderful and daring collection of unusual plants which Louisa cossets in the nursery and grows in beds and against walls; and the third, the towers. Three towers, a round gate-house, a fountain and bits of architectural ingenuity which are all James's, add height as well as a unique aspect to the garden.

In 1974 James laid out graph paper, pencil, ruler and rubber and for six or seven months worked at the design. 'It was a matter of getting the geometrical harmony so that it seems to be there naturally rather than that it had been dropped in.' He had little to go on except for fine brick walls eight or nine feet high around an asymmetric piece of flat land that had once been a vegetable garden. 'It was rather like a last piece of jigsaw the day it fitted, from every way you looked at it.'

There is a recurring theme of three routes in various parts of the garden that is most evident as you enter through the round gate-house, with its bell-tower. Three arches garlanded with roses, 'Veilchenblau', the colour of faded contusions, face you with alternatives. The central axis is strongly emphasized by wonderful tapering yew hedges that lead to a distant gate. To the extreme left is a small water garden contrived, designed and built by James. Here, needing shade and shelter, is *Mitraria coccinea*, a lovely thing with bright scarlet trumpet-shaped flowers, a *Desfontainea spinosa* recommended as suitable only in the home counties, advice the Arbuthnotts banish from their minds, and *Decumaria barbara*, a plant with small white flowers, another climber off-colour in winter weather.

To the right of the entrance is a bench where vigorous pairs of climbers intertwine, mingling their colours. Among some of the plantings, thriving in crowds, is a white rose, 'Aimée Vibert', supporting a strong, almost red clematis, 'Ernest Markham'. The reverberating combination of a crimson rose, 'Parkdirektor Riggers', tangles with the *texensis* hybrid, 'Duchess of Albany', one of the most ravishing clematis, with deep pink curling petals.

Stone House Cottage is not like other gardens. You don't want to approach it half-asleep, looking for drowsy romance and sensuous lethargy. Instead you need beady eyes, alert and appreciative, to take in a resourcefully planned design. Take care. You mustn't waste a step thinking of an overdue M.O.T. or what's for supper. In that second you will have missed an ingenious layout of one of James's crossing axes glimpsed between hedges, or Louisa's *Andromeda nana compacta* (of gardens) and *Itea ilicifolia* drooping poetically with fragrant flowers.

Louisa Arbuthnott is the most laid-back gardener I've ever met. Not only does she cope with four young children, some sheep, hens and ducks but she'll remark that, 'It's quite difficult to get two thousand plants into a fairly small space.' Oh yes? The suggestion has already made me feel peaky. 'I can't bear those great blocks of colour. I like itty-bitty effects. We just cram everything in!' (What did you say, Gertrude? Speak up a little.) Walk six inches and you'll see what Louisa means.

Like a series of cubby-holes, the individual gardens are each distinctly made for colour or for season. The first, surrounded by a box hedge, is for yellow and white plants such as *Trachelospermum asiaticum*, which grows against the wall with dark glossy leaves as background to its scented creamy-yellow flowers. Leading out from this is a broad shrub rose and herbaceous border jammed with colour, shapes and smells. Not a speck of uninhabited earth shows anywhere. Then further along the wall is another cubby-hole, the winter garden. This, too, is surrounded by a box hedge which shelters, amongst other things, *Abeliophyllum distichum*, which produces white flowers on leafless stems in February and is only suitable for the unflappable gardener as it's such a slow grower. Further on, against the first tower, is a climbing rose of such superlative splendour that any rose enthusiast would feel faint with longing to possess it. 'Cooper's Burmese', *Rosa cooperi*, is a prodigy of almost implausible beauty. Its single lustrous flowers, white and satiny, reach upwards against the mellow brick of one of James's towers. From this tower, on a fine evening in summer, the Arbuthnotts have a wind quartet wafting music out over the flower-beds while a conductor controls the musicians from another tower. Leading away from 'Cooper's Burmese' through an opening no wider than arms' breadth, is another axis crossing lawns to the far side of the garden. Turning back from there, through a doorway in the wall, is a small peat garden. Shady and secretive, there are plants here which thrive on acid soil. Things such as *Pieris japonica* 'Variegata', silvery and flushed with pink when young.

Around the house in raised or sunken beds are hellebores, hebes and a relation of the potato, *Fabiana imbricata* 'Prostrata', a mounded shrub with lavender flowers. Along the west wall where buttresses of evergreen *Pittosporum tenuifolium* give shelter and an architectural presence to Louisa's numerous climbers is one of that distinguished family of ceanothus, *C. arboreus* 'Trewithen Blue', a masterpiece. When it is in full animation the effect of dense blue fluffiness is a major event. Startling with its intensity, surely no one can withstand the temptation to try just once to grow this short-lived shrub? Along the west wall is a nice thing, *Drimys winteri*, a South American evergreen almost the darkness of olives with loosely formed ivory umbels.

In front of the pittosporum buttresses, lavish twiners, clingers, creepers and gropers, in front of the leathery leaves of tough survivors and the fiery colours of exotica rooted from tropical ancestors, is a huge spring garden bed of some of Louisa's two thousand plants. *Stachyurus chinensis* 'Magpie', with its translucent yellow flowers, is a spreading shrub of great distinction; so, too, is the yellow wood anemone, *A. ranunculoides*, a bit like a common wood anemone, growing with the mid-spring rosy-flowering currant, *Ribes sanguineum* 'Brocklebankii'.

You may visit this garden several times, moving from plant to plant at a snail's pace, and still miss a small enclosure within a sheltering yew hedge. A

secret garden: three or four little plots so enclosed and protected that a tender *Carpenteria californica* with immaculate single white petals likes it there, along with that most heavenly late-flowering shrub, *Hydrangea villosa*, which gave up on us years ago.

The garden is not all little enclosures stuffed with beds of flowers. There are areas to stride in, grassy spaces with dogwoods and berberis, elaeagnus and an odd legacy from David Douglas's travels in South America, a feathery flowering shrub, *Holodiscus discolor*. A crenellated hedge allows views across flat landscape to the distant misty mounds of the Malvern hills. Space and horizons didn't happen fortuitously, but through deliberation and intention in James's original design. To get an entire view of the garden, climb one of his towers. 'I've always been mad on bricklaying wherever I've been,' James says. Perhaps thinking of the winter of 1981/2, when the temperature went down to −26°C, he adds, 'There's something permanent about towers. And they have a functional purpose.' Tools are kept in one, with a telescope on the roof; a second is an adjunct to the nursery, where at ground level labelling is done and seed-drying above. The third tower has a balcony overlooking the garden and down below a wine cellar. 'I did feel awfully guilty building the towers when I should have been in the garden. But now we wonder what we'd do without them!'

Perseverance and boldness, not always available to gardeners, carry James and Louisa on imperturbably with their unusual and tender flowers. 'I love those delicate plants,' explains Louisa, 'probably because things that die half the time give me such a thrill when we have them back.'

From the beginning they had visualized Stone House Cottage as a garden for people to walk in. A place where visitors would be inspired, intrigued and helplessly tempted by what they had seen before walking into the nursery. The system is foolproof.

A third kind of garden was once everywhere, in the countryside, in villages and in small country towns: the cottage garden, the kind that lies like a garment close to the owner, almost an extension of their physical selves, certainly an extension of their love. Patios, flower-filled wheelbarrows or polystyrene Greek urns don't come into these places. Nothing is cerebral, designed or contrived; nothing is sophisticated, ingenious or trendy. The front path takes the shortest route to the front door, ignoring arty curves or devious effects. Shabby water-butts, relics of distant necessities, may still stand under deep eaves and mossy roofs; lop-sided wooden sheds for raffia and spades may still retain their practical lives in some corner of the garden, rather than as a support for exotic climbers. Damson and apple trees, long past their best, are cherished for their association, usefulness and antiquity. The true cottage garden is a conglomeration of nourishment – food for the stomach, flowers for the senses.

One such garden is Elizabeth Wheeler's. It reaches back into the past, not because she's been there long – she made it herself only ten years ago – but because she has brought with her, throughout her life, a memory of other places. A memory of her mother's and her great-aunt's gardens has been packed away somewhere into her seventy-six years, so that associations are carried as unbroken threads into the present. Primulas, honesty, wallflowers and penstemons cram the front garden. Tulips, with splodges of black in the centres, rose-red pulmonarias, hellebores and a 'Hardy Plumbago', *Ceratostigma willmottianum* – 'One of my mother's most favourite flowers – it's the most wonderful blue, the colour of gentians' – fill her five-by-eight-yard front garden. Pink astrantias, yellow asphodels – 'I saw them first in a garden about thirty years ago' – a woolly ballota (used in Greece to make wicks floating in oil for wayside shrines on country roads) and an alchemilla with silver-edged leaves are there.

'My rosemary died this winter but I bought a new one – it's such a lovely thing. Goodness, I do wish my pasque flower would grow a bit faster! Here's my latest extravagance, an arum lily that's supposed to be hardy', then with a bit of a giggle she adds, 'at least as far north as Pershore.'

We walk towards a small greenhouse looking as well integrated into the garden as does her old apple tree – 'It was an old tree when one of the past owners was a young girl. That was ninety years ago,' says Elizabeth. In the greenhouse is a whole variety of geraniums. 'My mother loved geraniums. She had such a collection!' Seedlings of stocks, larkspurs and mallows are growing in pots for a local flower festival in summer. Elizabeth points to another pot, 'These little bulbils I picked off a lily and they've taken so well. You know, I'm not sure whether they're strong enough yet to go out.'

Of all the plant associations with her past which infiltrate our wander round the garden, the most enduring is an agapanthus in a pot. 'It's much older than the hybrids. It must be more than a hundred years old. Well, it belonged to my great-aunt before it was my mother's.' For such an ancient plant it is flourishing. Five great blue flowers look likely to appear from its shiny strap-shaped leaves. 'I keep it in the same pot until it's really bursting out. They don't like being re-potted, it really sets them back.' Then with a deep sigh she adds, 'I love to have something belonging to my great-aunt because she was so lovely.' Other people might have hankered for a brooch or an ornament. 'On her fiftieth birthday she came down to breakfast with a proper cap on as a signal that now she was an old lady.' How surprised she'd have been to see her great-niece right now. 'It was such an adventure visiting her. We had a pony and trap to take us to the station. When we got out of the train my uncle met us with a dog cart, and the thing was to dash out of the station to get the dog cart over the bridge before the train started, or the pony would be all over the place. Oh dear, it was funny!'

At the back, where the garden is about the same size as at the front, Elizabeth remarks, 'I never wanted a lawn. I'd rather have flowers, as many flowers and as little paving and grass as possible.' Blackcurrants and red-currants, gooseberries, raspberries and rhubarb grow amongst viburnums, hollyhocks, sweet williams, forsythia, and a medlar. 'I've really given up the veg. garden. I've got a bit of a gammy knee and the one thing that upsets it is digging.' Her gammy knee may have been the end of the vegetable garden, but there isn't a piece of bare earth to be seen. 'Miss Wilmott's little rose (*Rosa willmottiae*) is there. That's a honeysuckle, I've forgotten what it is. This is a golden cut-leafed elder. A tree peony's coming there,' then laughing somewhat ruefully she adds, 'it hasn't ever blossomed yet.' Pointing to the back of the garden she remarks, 'The prickly thing beyond is acanthus and here's a little yellow poppy. Oh, and that's a lovely aquilegia. It's beautiful, but I have a fancy a white one will turn up in this mass somewhere.' We walk on a few steps, then stop as she looks down. 'I'm longing to see those irises. I got them from the West Midlands Iris Show. They'll be nice.' Against the wall is a clump of nettles, 'They're for the butterflies. And that's a wild rose. I'm wondering if I could do some budding. I don't really know quite how to set about it. Wouldn't it be exciting, but I think my cousin will show me.'

Elizabeth is a true gardener. It isn't just that her garden is filled with beloved treasures that she cares for so tenderly, but she never misses a meeting of the local horticultural society, she attends the W.I. plant sales and reads William Robinson. 'I've got my grandfather's encyclopaedia, the T.W. Sanders one. It's grand. It tells you the date of the plants and where they come from. I love it, I really do.'

Cottage gardens are an endangered species. Too much choice, too many varieties, too much hard-sell are thrown at us from garden centres. But as long as gardeners like Elizabeth exist, who grow flowers for their own sake and not to upstage the neighbours, then perhaps a few of this kind of gardener and garden, the antithesis of pedantry, may still be conserved like the blue butterfly or downy woundwort.

Wherever I go, whichever place I travel to, whatever garden I may finally sit in, I'll never be free of this one. Like some deep tenebrous scar I'll carry the making of our garden with me forever. It has been the first; the one from which so many garden thoughts have quickened, filling my mind with all those circles beyond circles of ever-widening ideas.

I know that making just one garden can last a lifetime. You can never draw a line and consider the garden to be finished. There are days when Michael and I walk through ours with gratification, others when we look with dispassionate eyes, when we see how much is yet to be done. In my garden book a trailing list of trees, shrubs and flowers wavers across the page. Names of plants we once urgently needed, fretting for instant effects, but which are now long since forgotten.

When we began, my most cherished book was Christopher Lloyd's *Shrubs and Trees for Small Gardens*, in which he gave a comforting list of things for 'Impatient Gardeners'. All that has passed. The frenzy is over. Avid plans, impetuous decisions when I longed by mere sleight-of-hand to have a dynamic garden, are no longer material. Thousands of plants surround us, deep-rooted and secure. The roses will perform in midsummer and the newts will return to the pond; each year the kingcups will appear all agog with yellow, and for the first time this spring the wild cherries in our wood have blossomed.

After nine years the pressure is off, the conflict is mollified and we can be less parochial and think of other things. Why, I've just read in the paper that a giant vegetarian dinosaur has been unearthed in Moreton-in-the-Marsh. I like that. I think I'll stop writing, see what Michael's doing and we might have a look at what was happening a hundred-and-sixty-five million years ago in the Cotswolds.

On 26 April 1989, since this book was written, Michael died.

Index

INDEX